T0110486

The Probability of God

THE PROBABILITY OF GOD

A Simple Calculation That Proves the Ultimate Truth

STEPHEN D. UNWIN, PH.D.

THREE RIVERS PRESS • NEW YORK

Published by Three Rivers Press, New York, New York.
Member of the Crown Publishing Group, a division of Random House, Inc.
www.crownpublishing.com

Originally published in hardcover by Crown Forum, a division of Random House, Inc., in 2003.

Library of Congress Cataloging-in-Publication Data
Unwin, Stephen D.
The probability of God : a simple calculation that proves the ultimate truth / Stephen D. Unwin.—1st ed.
Includes bibliographical references and index.
1. God—Proof. 2. Probabilities. 3. Bayes, Thomas. I. Title.
BT103.U59 2003
212' .1—dc21 2003008033

ISBN-13: 978-1-4000-5478-7

First Paperback Edition

144782292

To Heather

Contents

Acknowledgments

First and foremost I thank my wife, Heather, for her unfaltering support and confidence in all my endeavors—even the sensible ones. She was the first to critique each chapter you are about to read, and, invariably, her suggestions for improvement worked. I thank also my colleagues Robert Johnson and Steven Rudy, for enduring my relentless interrogation in the course of long business trips. Through their patience, they gave me invaluable insight into certain aspects of religious thought.

I am indebted to Kassie Rose, who was the first outside my family to read an early draft of the manuscript, for her encouragement and professional advice. I thank David Richardson at Prima Publishing for his immediate understanding of my project, his suggestions for improvement of the book, and his strong support throughout. Finally, my thanks go to Libby Larson of Prima for taking this first-time author through the book production process so competently.

one

Modest Objectives

Do you realize there is some probability that before you complete this sentence, you will be hoofed insensible by a wayward, miniature Mediterranean ass?

It seems you're one of the lucky ones. You were saved by an aspect of probability. Probability is the subject of this book: God, faith, and probability. I use the term *probability* in its strict mathematical sense and not in the fuzzy, ambiguous way it can be used in common language, as in "in all probability it's chicken." A mathematical probability is a number, and not just any number. It's a number that behaves in a strict, well-defined way, governed by the immutable laws of probability theory.

Furthermore, I mean the word *faith* in its religious sense: faith in God. The concepts of probability and faith are not often associated. One is of the domain of logic and analysis, the other of spirituality and religion. We are inclined generally to place a sharp divide between the two domains, and for good reason. At best, they tend to be mutually irrelevant, and more often, powerful mutual irritants. This is not to say that there has been any shortage of philosophical and scientific investigations on the reconciliation of reason with religious belief. This book is intended to contribute to that lack of shortage—not as a guarded, careful, scholarly work, however, but as a pragmatic, no-nonsense, bottom-line, results-oriented analysis. What I set as my objective is to calculate the numerical probability that God exists. Then, to pose a challenge, I will proceed to determine the relationship between this probability and the notion of religious faith.

Let me say first of all that I do not claim the ability to calculate this probability to an accuracy of, say, four significant digits, such as 81.91 percent. That would be an absurd claim given the extreme trickiness of the subject matter and would be legitimate grounds for tossing this book angrily into a fan. Two significant digits is the reasonable limit, such as 82 percent.

The concept of probability is germane to virtually every decision we make. When confronted by uncertainties, although we may not sit down and make a calculation of mathematical probability, we *do* rely on *probabilistic* thinking to cope with those uncertainties. Can I cross the track before the oncoming train arrives? Will XYZ stock gain this quarter? Will I be seen leaving before my boss's speech? In addressing such uncertainties, we typically consider the probabilities. A probability might be thought of as a value somewhere in a spectrum, perhaps defined in terms such as *low, medium,* and *high.* We then make a decision based on a combination of considerations, including the probabilities we have conceived. The acute importance of the question of faith in the existence of God warrants, in my mind, more than the fuzzy, vague thinking that we would expect to be employed by, say, a stock analyst or an oral hygienist. A more rigorous approach is demanded, as the following observation demonstrates.

What I set as my objective is to calculate the numerical probability that God exists.

Consider someone with a remaining life expectancy of, say, 40 years and a typical weekly church attendance duration of 1.5 hours. This person has a projected remaining church attendance duration of more than 3,000 hours! I understand for Baptists it may be much worse. This is not to mention midweek activities that may be influenced by the question of whether there is a God. Having realized this, I'm confident that nobody with a remaining life expectancy of 40 years would ask why this mathematical investigation is worthwhile. An accurate computation of the probability that God exists is surely a critical step in any day-to-day religious decision making.

Am I not daunted by the fact that the Greeks themselves, who created the foundations of modern science, mathematics, logic, and philosophy, were the first to abandon the notion that God could be understood within a logical, mathematical, or even philosophical framework? Am I not overwhelmed by the fact that the greatest minds in history have been unable to definitively answer the question of whether God exists but, in consolation, have produced incomprehensible treatises couched in impenetrable vocabulary, leading to vacuous conclusions? Well, yes, I am, but only a little, and here's why. These great thinkers did not think of addressing the issue of God's existence in a formal, probabilistic setting. They looked at the question in a strictly binary, deterministic way. They asked, "Is there a God, yes or no?" Some concluded yes (Anselm of Canterbury in the 11th century and René Descartes in the 17th century come to mind), and some concluded no (such as Arthur Schopenhauer in the 19th century and Bertrand Russell in the 20th century)—and for some, their conclusions, let alone their arguments, are not intended to be understood by anyone with a day job. I'm tackling this problem from a different perspective.

I'm starting from the premise that I don't know whether or not there is a God, just as the stock analyst doesn't know whether XYZ stock will gain or lose tomorrow. My objective is to calculate the *probability* that the true answer is yes, that God exists. Inherent in the language of probabilities is the ability to shy from dogma and from pronunciations of certainty (yet strangely, I will not always take advantage of this). I think this approach to the question is the one of intellectual humility.

In fairness, I should reveal the background and baggage I bring to the question. I started my career as a theoretical physicist. My area of research was one called quantum gravity, which I'll attempt to put in a nutshell. (This will transpire to be relevant.) An outstanding problem of modern physics is reconciliation of two major physical theories that have individually proven to be very successful descriptions of the world. One of these theories is Einstein's general theory of relativity, a theory of space, time, and gravity that provides a powerful description of the large-scale dynamics of the universe. The other is quantum theory, which provides an accurate description of physical phenomena at the microscopic level of molecules and smaller. These two theories, as is often the case with very successful individuals, don't like each other at all. If you try to account for gravity in quantum theory, the mathematics becomes ugly and unmanageable and does the kinds of things that make theoretical physicists wish they'd gone into dentistry. Attempts to reconcile these theories come under the rubric of quantum gravity.

This book is not about quantum theory or quantum gravity. The reason I mention them is because of their connection to the idea of probabilities. Physicists were bumbling along happily in the first part of the 20th century, thinking that they had just a few minor gaps to fill before they understood everything, when quantum theory came along and applied the intellectual equivalent of an electric whisk to their conception of the natural world. Suddenly, no longer was the universe considered deterministic (that is, cause A always results in effect B), but now the world was understood to be probabilistic (that is, cause A could result in a range of alternative effects, and all we can express is the probability of each). Albert Einstein didn't like this development at all (remember, he said, "God does not play dice"). Isaac Newton didn't even hear of it, since he was already long dead, but if he had, he would have claimed to have invented it. (This is from my pro-Leibniz joke repertoire and is not really relevant.) This probabilistic aspect of nature has some relationship to our current considerations, as you will see.

Back to me. I departed from the world of theoretical physics back in the 1980s and entered another area in which probabilistic thinking is paramount: risk analysis. I became a mathematical analyst of the risks associated with operating complex industrial facilities such as nuclear power plants, offshore oil platforms, hazardous waste repositories, and chemical plants. I learned how to state the risk of a major accident in numerical terms—that is, in terms of probabilities. Later, I'll go in some depth into what probabilities are, but for now, suffice to say that throughout my career, I've worked with them. So as you can imagine, it's a small step for me to embark on a calculation of the probability that God exists. Since the existence of God is the ultimate uncertainty and probabilistic analysis is the means of addressing uncertainties, so . . . well, it seems obvious: The probability of God begs to be computed.

So far I've confined revelation of my qualifications to the probability part of the *God, faith, and probability* question. What are my *God* qualifications? Well, to conduct this kind of analysis, I think that having any special religious or theological qualifications could be an enormous disadvantage. After all, if I were a bishop, a minister, a rabbi, or an imam, I doubt that you would be racked with anticipatory tension about the conclusion of my analysis. This is not to say that I lack religious knowledge. I consider myself to be the man-on-the-street when it comes to matters godly. Curiously, it might be argued that in this particular area, man-on-the-street knowledge does not necessarily fall short of expert knowledge. In fact, I'd go so far as to say that the God area is one that may have specialists but no experts.

After all, if I were a bishop, a minister, a rabbi, or an imam, I doubt that you would be racked with anticipatory tension about the conclusion of my analysis.

Like most of us, I was taught religion prior to the development of any intellectual defenses, and much of it stuck. I was raised in the north of England. My introduction to religion was the attendance of a Pentecostal church, of which my principal recollection is the speaking in tongues. This, the highlight of each three-hour Sunday service, was the time during which fellow churchgoers became seized by the spirit and would deliver a protracted oration in a locally unfamiliar language. That sort of thing would momentarily divert my attention from the Superboy comic book. Later in life I experienced the opposite end of the Protestant Christian spectrum. Beginning at age 11, I attended a school that was affiliated with an Anglican Cathedral. Moving from the ultraviolet to the infrared end of the Protestant spectrum, nobody spoke in tongues. Even English was strongly discouraged unless delivered in an acceptable accent.

Many years later, I moved to the United States, first serving as a minor diplomat and technical attaché. This was the title the British government gave me. I don't think it would have gained me diplomatic immunity for murder or other high crimes, but perhaps it was good for plea-copping lesser crimes such as small pet kidnapping. (I never tested this immunity.) I was stationed at a laboratory in New Mexico to play a small part in helping the governments of the United Kingdom and the United States see eye to eye on how to calculate the probability of a major nuclear power plant accident.

It was arriving in the United States almost 20 years ago that my religious sensibilities were shaken. I had become lolled into a somewhat lazy, complacent form of religious thought in which God had become simply a symbol of good ethical behavior, compassion, generosity, honesty, and other virtues. In a sense, the teachings of, say, Christ and Socrates had been, in my view, of the same character. Perhaps Plato, as the documenter of Socratic thought, had been a little more articulate and reasoned than Matthew, Mark, Luke, and John, but, nevertheless, I assessed the messages to be the same. I assumed that since the Socratic approach of stressing the intellectual imperative of ethical behavior had been an abject failure, the magic and hellfire elements of the Bible had been an alternative sales strategy, and a staggeringly successful one. Now, suddenly, I found myself in the company of people—educated and smart people—whose perception of God was far more concrete than mine.

To them, God was a supernatural entity, a person of sorts reflected by the fact that his demands upon us go beyond matters of *natural* morality. By natural morality I mean conforming to the Golden Rule, which is to "do unto others as you would have them do unto you." For example, he disfavors certain sexual preferences in people or prohibits the payment and receiving of interest on money loans or frowns upon the eating of certain comestibles. While it may be the case that any one person of faith believes that God has only some of these preferences, the idea is shared, nevertheless, that he has preferences at all. Indeed, these preferences, for people of faith, define moral behavior in a way that *natural* morality may not capture. So God, as viewed through religion, has personal tastes and preferences, like enjoying a good cup of tea but strongly discouraging the wanton consumption of coffee. (This particular example, to my knowledge, is not to be found in any sacred scripture and is presented as a hyperbolic analogy. Coffee drinkers may relax as far as they are able.)

So while in the back of my mind I had never abandoned the notion of God as a specific person-entity, as opposed to an abstract representation of the good, it was a shock to become aware of a mainstream perspective, residing in the *front* of peoples' minds, that a version of God similar to the one taught to me in Sunday school is real. Now, by no means were the perceptions of God in this new world uniform. Interpretation of the Bible, for example, provided one spectrum along which the Christian beliefs I encountered could be distinguished. While some think of the Bible as a history in the sense that it accurately relates historic settings, they exclude from its historicity the smaller-scale events and dialogues that can be construed as parables or events that clearly defy modern scientific understanding, such as the story of the creation and the occurrence of miracles. This view would make the Bible somewhat analogous to Homer's *Iliad* in that, while we may believe a great Bronze Age war took place on the Trojan plains, we view accounts of the battlefield participation of Ares, Athena, and other Olympian gods as poetic license. Others see the Bible as cover-to-cover literal history—more Thucydides than Homer—and inerrant to boot. Notwithstanding this kind of diversity in perspective, the commonality of belief in a real person-God is the striking part. So is there really a person-God? I got to thinking.

Whether God exists is a perplexing question that, if not properly dealt with, could easily consume every waking moment. That's why I wanted to get all this business firmly behind me. As I mentioned previously, the analysis and computations I will present are not intended to be scholarly, learned, theological assessments. This is simply a quick, pragmatic, but definitive analysis, intended to allow me, and any reader who agrees with my reasoning, to get on with life without having to continually revisit the issue. We can then incorporate the *probability of God* into everyday decisions with the comfort of knowing that we are behaving rationally, as the numbers dictate.

Summarizing, our objective is to calculate the probability that

God exists.

Further, it is to seek an understanding of the relationship between this probability and the notion of faith. Does the basis for faith increase as the probability of God increases, or is the relationship between the two far less straightforward? In setting the scene for this analysis, I will first need to define some terms and underlying concepts, such as *God*, *faith*, and *probability*. This I will do next.

Throughout this book I have referenced certain works. Rather than including distracting reference indices, I have simply listed reading materials in a bibliography, for those who are interested. The list is not intended to constitute an exhaustive catalog of relevant works but, rather, to provide a few appropriate starting points for further exploration of the disciplines and issues we will touch on. Finally, to help clarify where we are going and how far we have come as the analysis proceeds, key questions and conclusions are summarized at the beginning and end of each chapter. So off we go.

two

Not Just Any God

Question for This Chapter:

What shall we mean by the word *God* in our central proposition: God exists?

One beauty of working in the language of mathematics is that it forces us to nail down with some precision the exact nature of the problem to be solved. The reason that the impact of three millennia–worth of philosophical thought on everyday human life is about commensurate with the impact of the electric toothpick is that philosophers have worked largely in languages such as Greek, German, and English. These sorts of languages have proved very poor media in which to reason. They are completely porous to vagueness and ambiguity, which generally obviate the value of any conclusions reached. Indeed, it is the vagueness of common language that often constitutes its power, to which anyone who has written advertising copy can attest. Common language can endow the completely meaningless with an air of respectability merely through correctness of grammatical structure, such as "Flight 360 to Cincinnati is showing an on-time departure." Anyone who has flown understands the irrelevance of this proposition.

While it is never possible to entirely extricate oneself from the use of these common languages in problem solving, it is generally the case that the greater the reliance on the language of mathematics, the more confidence one is likely to place in the results. This is why it has proved utterly futile to express flight schedules, doctor appointment times, or party political manifestos in mathematical form. Math is altogether less fuzzy. (The notable exception to this rule is so-called Fuzzy Set Theory and Fuzzy Logic, invented in the 1960s by a fellow named Lotfi Zadeh, who clearly felt that mathematics had had, to that point, an unfair advantage.)

Before getting to the math, I want to take some time to clearly define the subject of our analysis to the extent allowed by the aforementioned limitations of fuzzy language. Our objective is to calculate the probability that the following proposition is true:

God exists.

Henceforth, let's refer to this as Proposition G. (This should not be confused with the sort of Proposition G that is the subject of local government referenda, as in "Vote Yes to Proposition G—the Second Amendment applies to grade school teachers too.") Now, as stated, Proposition G is exactly the sort of fuzzy, vague sentence that I just went out of my way to criticize. The mathematical system of probabilities, if it had voice, would utterly reject Proposition G as a viable component of analysis unless it were qualified in a way that endowed it with meaning. Put another way, a probability cannot be attached to a meaningless proposition. For example, take the proposition

I did not have sexual relations with that woman.

If, hypothetically, someone made this statement and you then asked me for the probability that it's true, I would have to ask what was meant by "sexual relations" before I could even begin to address the matter. Without that definition, the idea that the proposition is true or false is meaningless, and, therefore, meaningless also is the question of the appropriate probability that it is true. You might as well ask what the probability is that the proposition that the Pacific onion beholds moral valence is true.

Back to Proposition G. The word *God* can be used in a broad range of contexts. If we do not try to narrow its meaning for our current purposes, then I doubt that any probability we calculate could be of much interest to anyone. Unless we are more specific about God, it's arguable that the probability of the proposition being true is automatically 100 percent since the word can be used in some very weak senses, such as "God is the universe" or "God is love." If either of these definitions of God were employed, then to derive a truth probability for the proposition that is less than 100 percent would be to doubt the very existence of the universe or of love. While I wouldn't put those doubts past some people (solipsists and Germans, respectively), I won't waste *your* time with such extreme cynicism.

As a recovering theoretical physicist, I would like to begin with consideration of Albert Einstein. Professor Einstein was prone to allude to God more than the average physicist. I've already quoted his famous "God does not play dice" observation in the context of his concerns about quantum theory. Another of his remarks was "God is subtle but he is not malicious." Einstein was never reluctant to criticize new physical theories that violated his idea of God. Take, for example, a theory of space, time, and matter proposed by the German mathematician Hermann Weyl. In this theory, the physical size of an object depended upon its location in space such that, for example, a 6-foot man only becomes 6 feet tall when his location coincides with that of the tape measure. If he moves to a different location, he could be the size of a bread box or of a galaxy. However, if he took the tape measure with him, it would undergo the same stretching or shrinking experience as that undergone by the man and thus would still measure him as 6 feet. That is, whereas a rose is just a rose, a foot is not always a foot. Incidentally, this is precisely the sort of thing that happens when pure mathematicians are allowed to do physics. Anyway, Einstein was thoroughly unimpressed by this theory, and by many others also. They failed to meet his expectation of God's universe. So when Einstein invoked God in his opinion of a physical idea, what did he mean by it? Was he referring to the same person-God we discussed earlier, who may like tea but not coffee?

"No" is the answer. Einstein was a disciple of the 17th-century Dutch philosopher Baruch Spinoza. The Spinozan concept of God is quite different from that of the person-God of the major monotheistic religions. For Spinoza, God was indistinguishable from the natural universe. To him, nature was the physical manifestation of God. God is all things and all things are God, or to quote Spinoza himself, "Nothing exists but God." This is the so-called pantheistic view. In this view, God is not someone to be interacted with personally through prayer or other means or someone who influences the courses of our lives or hosts any kind of afterlife events. He is simply all things. He *is* nature. This is the antithesis of the person-God view that revolves around God as the master of supernature, as a standalone entity who influences nature from the outside.

Professor Einstein was prone to allude to God more than the average physicist.

That Einstein was of the Spinozan view is particularly poignant. He saw the physical elegance and beauty of nature as the realization of God. That the world can be reduced to simple but subtle physical principles and, furthermore, can be understood in terms of human mathematics and logic was revered by Einstein. He said, "When the solution is simple, God is answering." The elegance and beauty of nature was something about which Einstein felt deeply, and he was quick to criticize any suggested physical theories that violated this principle. Incidentally, even though Einstein had great problems with the concepts of quantum theory, most physicists are now of the opinion that this theory is quite consistent with the elegance principle. I'll give you a final quote on the matter, coming, ironically, from the mathematician Weyl, whom I just maligned. He said, "God exists since mathematics is consistent, and the devil exists since its consistency cannot be proved." Maybe the old German got it after all.

So while this Spinozan viewpoint is one way of thinking about God, it is generally not what you hear in church, the mosque, or the synagogue. This sort of thing would do no good in collection-plate terms. Few would be eager to spend their Sabbath praising the ability to conceptualize gravity through differential geometry. This kind of application of the word *God* is all well and good, but it couldn't be more dissociated from the religious person-God. Good versus evil, heaven versus hell, compassion ver-

Few would be eager to spend their Sabbath praising the ability to conceptualize gravity through differential geometry.

sus indifference, personal communication through prayer, being sustained and protected, and justice and mercy—these are the elements of belief in the person-God. The reason I have taken this brief diversion into the God of pantheism is to stress that this, and other philosophical perspectives on theism, are *not* the notions of God that our proposition addresses. The person-God of the major faiths is the subject of our analysis and *not* the God of Spinoza, Einstein, or any other fancy philosophical school. Our Proposition G refers to the God of Christians, the Jehovah of Jews, the Allah of Muslims, the Wise Lord of Zoroastrians, et cetera. Although there is some disagreement between and within religions about the specific characteristics of the person-God, the similarities in beliefs outnumber the differences. Put another way, followers of these religions could be relied upon to gang up on any hapless pantheist who found herself in the wrong part of town.

Having nailed down the God element to the degree necessary for progress, let's take a brief, preliminary look at the word *faith,* a concept we will come to explore in greater detail later in our analysis.

I think it fair to say that a loving, compassionate God who keeps to himself would not be critically relevant to our lives. A passive God whose activities are confined to empathy with our hardships but who declines to effect any influence is not the person-God of the major faiths. People of faith have varying beliefs about the mechanics and timing by which God exerts his influence, but they agree that influence he does. This influence is often conceptualized in terms of just but merciful outcomes to all our earthly predicaments. One school of thought is that God ensures just outcomes right here on Earth. While "God works in mysterious ways" that we may not always comprehend, the outcome is always the *right* one. Another school, more consistent with Python (Monty) Eric Idle's lyric "Life's a piece of shit, when you look at it," argues against this notion of earthly justice, believing that it is in an afterlife that all wrongs are righted. Some have it that our fate in the afterlife is determined less by our worldly deeds but rather by the religious beliefs we held here on Earth. All of these ideas share the notion of divine, just, and merciful outcome tailored to each human being during life and afterlife. Belief in a God with this disposition and in the *correctness* of all ultimate outcomes is, in my view, the stuff of *faith.* Or at least this is a working start for our deliberations and analysis. More about faith later.

Chapter Conclusion:

We will use the word *God* in the traditional sense associated with the major monotheistic faiths and not in a more philosophical, abstract way.

---

three

You Are Here

Question for This Chapter:

Do the general cosmology, physics, and biology of our universe constitute relevant evidence in assessing the probability that Proposition G is true?

In chapter 2, I identified the type of God that is *not* the subject of this analysis. In that same negative vein, I will now identify the kinds of arguments and logic that we will *not* be using in the determination of the probability of God. This is worth doing right at the outset since there seem to be certain pervasive themes in the assembly of evidence for God that we need to address and dismiss. I am referring principally to the notion that could be articulated as follows:

> This couldn't have all just happened randomly. Look at the complexity in the world. Look at the intricacy and beauty

> of a rose petal and the miracle of a newborn baby. There must be some type of intelligent design behind all of this. It can't be the outcome of random chance and the indifferent processes of nature.

Physicists have every reason to empathize with this viewpoint since they know how exceedingly unlikely it was to have ended up with this universe we inhabit. We know that systems enjoy chaos far more than order, and, given the choice, the universe would much sooner have selected chaos for itself than the strict disciplinarian existence it actually leads. That is, there were so many more chaotic options than ordered ones for the universe to choose from that random selection would, with overwhelming probability, have resulted in the choice of chaos: universal primordial soup rather than ordered rose petals. We would never expect, for example, a box of Lego bricks thrown into the air to fall by chance into a model of the Château de Versailles or for my son's bedroom to become tidier without the imposition of some miraculous force. Therefore, if the universe were formed through purely random processes, we should expect it to comprise nothing but chaos and not, as we see, galaxies, stars, and shopping malls. In the language of physicists, the specifics of the initial state of the universe in combination with the nature of the physical laws that governed its subsequent evolution could easily give the appearance of having intelligently conspired to produce an ordered, life-bearing universe.

Even this expression of the unlikeliness of it all understates the miraculousness of our ordered universe. Going back to the analogy of the tossed Lego bricks, imagine that the stud and tube couplings in the bricks had not intentionally been spaced and shaped to bind each other but had been randomly positioned. This would make the random formation of the Château de Versailles even more staggeringly surprising. The analogy is that the physical world produces certain so-called fundamental constants of nature that just happen to have the values they have because . . . well . . . just because they do. These include, for example, the gravitational constant of Newton and Einstein, the speed of light, the electric charge of an electron, the mass of a proton, a constant from the quantum world named for Max Planck, and others. These fundamental constants in a sense calibrate the laws of physics. Now, had the values of these fundamental constants been different by just a few percentage points, then the repercussions would have been enormous: Galaxies would not have condensed from the primordial soup, then stars, planets, and ultimately human life would not have appeared. Yet those constants are seemingly fine-tuned for order, and galaxies, stars, planets, we humans, and our pets are the consequence. So don't think for a minute that physicists fail to understand the rose petal argument for God.

We would never expect, for example, a box of Lego bricks thrown into the air to fall by chance into a model of the Château de Versailles.

They are seized, no less than anyone, by the strangeness of it all.

Notwithstanding this strangeness, I do not see the rose petal matter as an element of our analysis and of the question of the probability of God. To explain why not, I'd like to begin by presenting something I call "The Shopping Mall Vignette."

THE SHOPPING MALL VIGNETTE

Scene: A shopping mall.
Characters: Two teenage boys—Anaxagoras (The Axe) and Chad.

THE AXE: We're lost.

CHAD: I know. I think society has let us down by failing to instill in us a sense of those values that . . .

THE AXE: No, you idiot. I mean it literally. I've no idea where the Body-Part Piercing Parlor is.

CHAD: Oh. There's a mall guide over there.

[CHAD *walks over to the mall guide, which is a wall map showing the locations of the mall stores. He inspects it for a while.*]

CHAD: Oh, my god. [CHAD *is unnerved.*] This is totally incredible. I just don't believe it.

THE AXE: What is it? Speak to me, Chad.

CHAD: Come and look. You won't believe this.

[THE AXE *joins* CHAD *and looks at the mall guide.*]

THE AXE: Well?

CHAD: Don't you see it? It took me a while to notice it too.

THE AXE: Notice what?

[CHAD *points to an arrow on the map beside which is written "You Are Here."*]

CHAD: We *are* exactly where the arrow says we are. How could they have known that?

[THE AXE *ponders the situation.*]

CHAD: I knew that these shopping mall people were clever, but this is miraculous. [CHAD *looks up suspiciously for cameras.*] [*Whispering.*] How do they do that? How did they know where we are? Axe, I'm scared.

THE AXE: Let's think this through. I don't believe in miracles, not even in a shopping mall.

CHAD: I'm listening. [CHAD *is still surveying his surroundings suspiciously.*]

THE AXE: Say that we were standing over there by the Nasty Jewelry Cart. You can't read this mall guide from there. The guide would still assert that "You Are Here," but it would be wrong. However, we wouldn't be reading the mall guide if we were over there, and, therefore, we would be unaware that the guide's assertion is false.

CHAD: Over there by the Nasty Jewelry Cart?

THE AXE: Yes. So only when we are able to read the guide is the guide true. The guide would never be read and found to be false. Any perceiver of the guide would find it to be true by virtue of the necessary attributes of the perceiver—that is, his location. It has nothing to do with omniscience of the mall management.

CHAD: [*Absentmindedly, staring at the Nasty Jewelry Cart.*] So where's the Body-Part Piercing Parlor?

Curtain falls.

This is the analogy: Only an ordered universe can be observed. That is, only in a universe that has the correct conditions for life would that universe be perceived. In no hypothetical universe would the following observation be overheard: "Just as I thought, no life here." Just as the "You Are Here" arrow is guaranteed to be correct for anyone who is reading the mall guide, so a perceived universe must be ordered. This way of thinking is sometimes referred to as the *anthropic principle.* The only meaningful universe is a perceived one, and to be perceived it must have those attributes necessary to produce life, just as the mall boys required the attribute of being next to the mall guide in order to note its correctness in the assertion "You Are Here." Therefore, surprise at the existence of life, like surprise at the correctness of the mall guide, is unwarranted.

You may not like this argument and may retort, "But surely it was overwhelmingly likely that there would be nobody around and that nobody would be the occupant of a chaotic universe; that is, the universe would be lifeless." Likewise, in the shopping mall there was a legitimate possibility that The Axe and Chad might never have been in the vicinity of the mall guide, and so the question of the guide's accuracy would not have even arisen. Let's delve further into this, but now with a little more emphasis on the idea of probabilities.

First, I want to ponder the notion of the a priori odds of an event occurring and how those odds relate to the appropriate level of surprise if that event then actually occurs. Say you pick the top card from a shuffled deck. You look at the card and it's the jack of spades. Are you surprised? Probably not—why *shouldn't* it have been the jack of spades? Well, I think it would have been a whole different matter had someone first asked you about the odds of picking the jack of spades, and *then* you drew the jack. At 51 to 1 against, you would have beaten some pretty slim odds. In fact, I suspect that your degree of surprise would have been in proportion to the a priori odds against the event. Imagine now that all alternative universes were represented by a cosmological deck of cards, and the one that was randomly selected was our universe. It is true that the vast majority of cards in the cosmological deck represented chaotic universes incapable of sustaining life, just as 51 cards had the attribute of *not* being the jack of spades. So why should we be surprised about our ordered universe if you're not surprised about having drawn the jack? No miraculous mechanism should be suspected.

On the unlikely premise that you are still uncomfortable with this argument, let me now try to tie together the jack of spades perspective with the anthropic principle rediscovered by Chad and Anaxagoras, the mall boys.

Hypothesize a multitude of parallel universes. (Yes, this will be a down-to-earth, pragmatic argument as promised.) They represent all possible physical universes, and so, we assume, the vast majority of them are chaotic and incapable of sustaining life; that is, most are equivalent to a scattered array of Lego bricks and *not* to the Château de Versailles. Imagine, for instance, strange parallel universes in which teachers and fire fighters become wealthy, and it's the lawyers and physicians who have to settle for adulation, or even stranger universes with watchable television. In this thought experiment, all possible universes *do* exist, and it is not surprising that the one we inhabit and observe is one of the ones that *can* sustain life. After all, I am not surprised that I live in the American Midwest rather than in the vacuum of space, even though the latter is tremendously more abundant. This perspective seems to suck the probabilistic awe from the situation.

Interestingly, in the 1950s the physicist Hugh Everett suggested that the world of quantum theory could be understood in terms of multiple, parallel universes. His theory has it that these universes are continually branching away from each other, reflecting the range of possible outcomes associated with the a priori uncertain behavior of quantum systems. In this multiverse model, the more recently two universes branched away from one another, the more similar those two universes are. The multiverse model would even accommodate a range of alternative initial big bang conditions, reflecting the possible outcomes of a primordial quantum state. Now, these speculated universes could be dismissed as philosophical fluff if it were the case that there existed no means of interuniverse communication. After all, if the existence of a parallel universe is unknowable, then we are completely unconstrained in our speculation about its specific properties, and so such speculation is worthless. However, these Everett universes *do* communicate with each other to some limited degree. Phenomena such as optical interference patterns are understood in the Everett model in terms of interactions between the alternative paths taken by a photon of light in each parallel universe. This Everett idea may be viewed as bringing some sense of physical solidity to our anthropic argument. Indeed, this multiverse model seems to be in current favor among physicists as the basis for interpreting some of the strange implications of quantum mechanics.

Before we move on, a variant of this rose petal argument for God is worth mention. This variant maintains that there are certain biological phenomena that cannot be explained by modern evolutionary biology. For example, some claim that there is no evolutionary mechanism for the emergence of the eye. Their position is that there are no scientific, evolutionary mechanisms that could explain this complex, biological phenomenon, and so this can only be the work of God. The term *intelligent design* is sometimes used to express this notion by those who feel a little queasy about direct reference to God. In a sense, this is the most primitive notion of God—God as the explanation for gaps in scientific knowledge. In ancient times, these gap-phenomena included sunrises, thunderstorms, earthquakes, and the like; now they are sophisticated biological phenomena. (Incidentally, most of these claims about the inability of evolutionary mechanisms to account for certain complex biological systems have, to my knowledge, been shot down in flames—but that's beside the point.) The unsettling aspect of this notion of God as a gap-filler in understanding is that as scientific knowledge increases, God inevitably shrinks. In this model, current ignorance is the fuel for belief in God. This cannot be an appropriate role for him. This idea of God is, in some respects, the very antithesis of the Spinozan God who represents the revealed beauty and elegance of nature.

My humble view is that if we wish to ascribe to God the creation of the very laws of nature, then I suspect that those laws would be sufficiently perfect to result in any physical universe he wished for, and there would be little reason for any subsequent tinkering on his part that would give us the impression of extra-natural design. That is, whether or not God created natural law, natural law does the job perfectly well. Now, some view intelligent design not in terms of divine, postcreation tinkering but rather as the design of the physical laws themselves in a way that ensured a life-producing universe. This may be a stronger argument. However, we might counter by extending our multiple universes conjecture to capture not only alternative world evolutions that are subject to our laws of physics but also alternative sets of the actual physical laws themselves. So, for example, in some of these universes, like magnetic poles would attract one another and opposite poles would repel. Then we could apply similar arguments as before and deduce that, obviously, we live in one of the universes whose physical laws allow us to exist. Thus the anthropic principle again prevails.

The unsettling aspect of this notion of God as a gap-filler in understanding is that as scientific knowledge increases, God inevitably shrinks.

Well, it might be argued that this bold expansion of the multiple universes idea is a little spurious since, by definition, it cannot be hypothesized within the constraints of the physical laws we humans took so long to uncover. Furthermore, we seem to lack the quantum mechanical justification that applies to the more limited multiverse model. In effect, we throw open the doors to the possibility of other sets of laws just to rationalize why *our* laws are conducive to life. But the other side to that coin is the question of whether we *need* to speculate on alternative sets of physical laws or whether there is some logical necessity constraining physics to be precisely as it is.

There have been recent speculations that there may exist logical, physical reasons why our universe was forced to turn out as highly ordered as it did. These speculations have it that some underlying (although as yet undiscovered) supertheory of everything might render far more likely than we had originally believed the outcome of a highly ordered, life-sustaining world. So, for example, while we now marvel that the fundamental constants of nature (such as the speed of light, the mass of an electron, and the strength of gravity) seem to be precisely tuned in value to allow order and life, it may be the case that these constants are not so fundamental after all and their precise values are in fact *necessitated* by more profound physical principles. For example, the so-called M-Theory is one that some physicists believe has the potential to become an embryonic version of such a supertheory. These types of theories hold out the possibility of a physical framework in which we can understand the suppression of chaos, and without recourse to anthropic arguments. Paul Davies provides a lucid exploration of such speculations in his book *The Mind of God* (1993).

In conclusion, I do not believe that the rose petal argument is a good one for God. I'm not suggesting that any of this anthropic, probabilistic thinking disfavors the existence of God, just that it is not the sort of evidence that points in either direction, for or against. I do concede that at some future time there may exist a consensus among scientists and philosophers that we do indeed need to resort to some as yet unknown profound principle or idea to understand the life-bearing nature of our universe. Perhaps those scientists and philosophers will even be so bold and arbitrary as to attach the word *God* to that principle. Yet even in this hypothetical scenario, association of *that* God with the person-God of *our* analysis would be quite a leap since a mere explanation of physical and biological order would unlikely influence the probability of the compassionate, caring, and personal God known to the major faiths.

I believe that any science-based argument for or against the existence of the person-God is troublesome. Looking for God in science is like breaking open a television set to look for the tiny actors inside. The phenomenon of television can be understood either in terms of cathode ray tubes or in terms of TV shows. These understandings are equally valid but mutually irrelevant. The continued existence of debate involving both science and religion is very depressing, proving that we as a species have not come that far. Of course, if human ignorance fuels that debate, then we are cursed with an inexhaustible energy supply. As a tangential consideration, imagine that we could use ignorance as automobile fuel. We would never need to import another ounce of oil. In fact, we would have huge energy surpluses but nowhere to send it. Could Texas maintain its preeminent position as a natural energy resource?

I would conclude that the question of God's existence cannot be addressed through reference to cosmology, biology, or any other science. To plagiarize and adapt from the best: Render unto the physical world those things that are physical, and render unto God those things that are God's.

Looking for God in science is like breaking open a television set to look for the tiny actors inside.

To backtrack immediately on this whole rendering idea, there is clearly one scientific knowledge gap that may lack even the potential to be filled and thus provides a God-slot, and it is the question of why *anything* exists. Even here, there exist quantum physical models developed in the 1980s of how the big bang occurred. For example, there is the so-called inflationary universe model in which the big bang is the result of a quantum fluctuation that got carried away! Of course, the model relies on the laws of physics, and, as we discussed, we can ask who created *them*? I suppose that creation of the very principles of nature is the most credible God-slot. In Aristotelian reasoning, we must consider each event in the world to have a cause, and that cause to be an event that has its own cause, et cetera; but eventually we must reach a brick wall of causality. This brick wall represents the *original* cause effected by the Prime Mover. This Prime Mover is Aristotle's God. But is this God just another philosopher's God (for if Aristotle's God isn't a philosopher's God, who is?), produced from a handkerchief by the application of clever logic? Would this logic necessarily imply the existence of our caring person-God? Again, a God invoked by clever logic or science does not lead us necessarily to the God of faith.

Anyway, in the pragmatic spirit of our analysis, let's get on with it and not wait around until these philosophical and scientific matters are resolved. So now that we've dismissed the relevance of the universe in our considerations, let's proceed.

Chapter Conclusion:

The physical and biological laws of our universe and the phenomena to which they give rise do not provide meaningful evidence either for or against the proposition that God exists.

four

The Good Reverend Bayes

Questions for This Chapter:

What is probability?

What is Bayes' theorem of probabilities?

How is Bayes' theorem generally applied in coping with uncertainties?

We will now grapple with the notion of probability. First, I stress that I mean the word *probability* in a mathematical sense and not in some everyday, vague, qualitative sense. When used in common language, the word can have a range of meanings. If a trial lawyer uses the word in the context of a lie such as "This evidence, members of the jury, surely shows the *probability* is that my client was nowhere near the fruit stand," he means that his client was *most likely* not at the scene in question. Using the word in a different sense, a physician might observe, "There is little *probability* that it got there by accident." In this case, the word is used as a qualitative scale of likelihood. In a world where language is evolving faster than Darwin could feed a banana to his uncle, I'm confident that by now the word *probability* has numerous additional applications. In conclusion, while common language may be useful for ordering a corn dog or speculating with a telecom telemarketer on the subject of his parentage, its value in logical analysis can be limited. For that you need the language of mathematics.

In contrast to fuzzy linguistic probabilities, mathematical probabilities have very clearly defined properties. They behave in ways that are strictly governed by the mathematical theorems of probability. To begin with, a number representing a probability must lie in the range of 0 to 1 or, if stated in percentage terms, in the range of 0 percent to 100 percent. A typical articulation of probability is in the form "I'd say we have a 50-50 shot," referring to a 50 percent probability of the desirable outcome and a 50 percent probability of the opposite.

The most widely understood notion of mathematical probability relates to the idea of event frequencies. When we say a fair coin has a 50 percent probability of landing heads up, we mean that if we repeatedly tossed the coin, then in the long run it would land heads up 50 percent of the time. A similar interpretation applies to the statement that a fair die has about a 16.7 percent probability (one-sixth probability) of landing three dots up. This is the so-called frequentist interpretation of probabilities, where a probability number is equal to a statistical frequency of some class of events, such as coins landing heads up. However, if this frequentist idea were strictly adhered to, then the applications of probability theory would be very limited, as will be explained.

Consider the price per share of Monopolistic Practices, Incorporated, with ticker symbol MPI. At the opening of the stock market, you ask your stock adviser for his assessment of the probability that MPI stock will have gained by the close of market. Could your stock broker (once he had finished laughing) answer this question based on the frequentist interpretation of probabilities? Well, if he were not too bright, he could look at the history of the stock and consider the fraction of days on which the stock has gained and call that fraction the probability that the stock will gain today. More realistically, however, the types of factors he would consider in arriving at the probability of a gain would be the anticipated content of the forthcoming MPI earnings report, the overall strength of MPI's industry and customer sectors, the performance of the stock market overall, and other fundamental determinants. So when he replies, say, that there is a 75 percent probability of an MPI gain today, this number really has no frequentist interpretation. He did not derive it by thinking about repeated events such as the tossing of a coin.

To take a more extreme example of a situation in which traditional frequentist thinking has no role, consider the case of the neutrino, a fundamental particle of matter. Currently, there exists uncertainty about whether neutrinos are completely without mass or whether they have some very small but finite mass. You may ask a physicist whether she believes that neutrinos are strictly massless, and she may reply, "There's a 5 percent probability that neutrinos are massless." What does she mean by this? She certainly doesn't mean that 5 percent of neutrinos are massless since, as a class of particles, they either are *all* massless or *all* have mass. Those coin-tossing, die-throwing frequentists would be utterly bewildered by her statement of probability. I daresay that their bewilderment would pale next to the synapse-bursting shock of discovering that our mission is to calculate the probability of Proposition G. Well then, what *did* the physicist mean?

Between the 1920s and 1950s, a school of mathematicians, most notably Frank Ramsey, Bruno de Finetti, and Leonard Savage, developed an interpretation of probability theory based on the recognition that the frequentists' view was very limiting, as the hypothetical stock adviser and physicist would agree. They sought to redefine the way in which we think about mathematical probabilities while retaining the strict mathematical formalism of probability theory. Their work has resulted in what we call today the Bayesian interpretation of probabilities. You might have expected the name of this new interpretation to be taken from its originators so that we would have Savage probabilities (a good movie title, by the way) or de Finettibilities, but no. I'll explain the origin of the term *Bayesian probabilities* shortly.

In the Bayesian world, a probability is an expression of a degree of belief. When the statement is made by the physicist that there is a 5 percent probability neutrinos are massless, she means that on a scale of 0 to 100, her degree of belief that the said particles are massless is 5. You may ask, "But isn't this all reverting to the fuzzy arena of linguistic probabilities?" Well, no. The beauty of casting degrees of belief in this probabilistic framework is that probability theory then dictates strict, mathematical means of deriving quantitative probabilities based on the evidence that is available. In particular, Bayes' theorem, one of those basic theorems I referred to earlier that govern the mathematics of probabilities, is a critical ingredient in the computation of Bayesian probabilities.

In the Bayesian world, a probability is an expression of a degree of belief.

Given the subject of our analysis, it is somewhat pithy that Thomas Bayes, the man for whom the theorem was named, was an 18th-century English Presbyterian minister. However, it was not until de Finetti and Savage got into the picture that the modern interpretation and full relevance of this theorem became established. Before waxing theoretical about the theorem, let's briefly return to the frequentists and not just leave them on the roadside like a colony of flat squirrels. Frequentist notions are still of great value. Nobody can deny that associated with the toss of a fair coin is a 50 percent probability of it landing heads up. However, the way a Bayesian would describe the situation is as follows. Instead of thinking of the statement "The probability of heads is 50 percent" to be simply a reexpression of the statement "The coin lands heads up 50 percent of the time in the long run," think about it this way: If you ask what the probability is that the following proposition is true:

> The next toss of this fair coin will results in heads

then given that 50 percent of tosses result in heads, a rational person would adopt that frequency as his degree of belief, or probability, that the proposition in question is true (i.e., that the next toss will land heads up). This may seem like a subtle distinction: A statistical frequency of an event, when known, is *adopted* by a rational person as a probability, as opposed to being *defined* as a probability, but it is an important distinction since most propositions do not have the luxury of being related to statistical processes like coin tossing. We have to think of the rights of these disadvantaged propositions. This Bayesian approach, in principle, allows probabilities to be attached to a much broader range of propositions, such as our Proposition G.

So what is this mathematical theorem derived by the good Reverend Bayes, and why is it so important that his name has been posthumously attached to an entire philosophy of probability? First, it is important to understand that the fundamental value of probability theory lies in its capability to provide the basis for representing uncertainties in mathematical form. When you see reference to probabilities, you are invariably looking at a situation where uncertainties exist. When we refer to the probability of a particular outcome to a situation, there must be uncertainty as to what that outcome will be. Without uncertainties, there is no role for probabilities. Now, it is an unfortunate yet adrenalin-producing aspect of life that uncertainties are rife and thus most decisions are made under some degree of uncertainty. Articulations of such decisions might include "Buy 1,000 shares of MPI," "We find the defendant guilty as charged," "Yes, we can go out for a drink sometime," and "Put him under, we're going in." In each of these cases, the evidence was weighed up, probabilities were assessed (perhaps explicitly through mathematical methods or perhaps more intuitively), and a decision was made. Each decision was made under uncertainties, but, nevertheless, a decision had to be made, and the best that could be done was to weigh the evidence, see how it affected the probabilities, and then proceed. The reason that Bayes' theorem is so important is that it provides the mathematical framework for dictating the

It is an unfortunate yet adrenalin-producing aspect of life that uncertainties are rife and thus most decisions are made under some degree of uncertainty.

way in which evidence should affect the assessment of numerical probability.

It's about time to bite the bullet and express Bayes' theorem in its full mathematical glory, but first a few preliminaries regarding notation. Mathematics demands the use of abbreviations; otherwise, the simplest formulas would be interminable, and entire equatorial rainforests would have to be destroyed just to meet the paper needs of a 9th-grade math student. For example, who could argue for the expression

> Energy of object at rest equals rest mass of object multiplied by the speed of light in vacuum to the power of two

instead of its elegant abbreviation $E = mc^2$? In that spirit, we will abbreviate our proposition

> Proposition G = God exists

to the symbol G. That is, when we write G, it is shorthand for the proposition that God exists.

Now let's introduce the idea of the negation of a proposition. The negation of a proposition is, in a sense, the opposite proposition. For our Proposition G, its negation is denoted G*. G* is then shorthand for the proposition that God does *not* exist. G and G* together are, in mathematical terminology, a *mutually exclusive* and *exhaustive* set of propositions. They are called mutually exclusive because G and G* cannot both be true; that is, if one of these propositions is true, then the other is false. They are called exhaustive because G and G*, in combination, cover the entire gamut of possibilities—there is no third option. To relate these propositions to the archetypical religious perspectives:

Theists would say:	G is true and G* is false.
Atheists would say:	G is false and G* is true.
Agnostics would say:	Don't know whether it is G or G* that is true.

Already mathematics is imposing a certain cleanliness on the situation, don't you think? It seems analogous to the physician who has no clue what is wrong with you, but he has a name for it.

Now we introduce probabilities. First, there is the term P(G):

P(G) means the probability that G is true.

Let us presume that on some basis we had determined the value of this probability. However, say that some new evidence now comes along that bears on the issue of whether G is true. Having fully surrendered to the abstract world of mathematics, let's represent this new evidence by the symbol E. How does this new evidence affect our assessment of the probability that G is true? This is where Bayes' theorem comes in. First, let's establish a notation for the postevidence probability:

P(G|E) means the probability that G is true after the new evidence is accounted for.

So, in effect, we have the *before and after* probability: the probability assessed prior to the new evidence and the probability assessed after the new evidence is considered. In Bayesian jargon, these are known as the *prior* probability and the *posterior* probability, respectively. P(G) is a prior probability, and P(G|E) is a posterior probability. Proposition G*, that God does not exist, also has its prior and posterior probabilities, denoted P(G*) and P(G*|E), respectively. At this point, I speak for the entire Bayesian community in denying categorically that the term *posterior probability* refers in any way to the part of the human anatomy from which such probability estimates are produced. That is merely a foul rumor spread by those coin-tossing frequentists I warned you about.

There is one more type of probability I need to explain before getting to Reverend Bayes. Just as P(G|E) means the probability that G is true given the evidence E, can you guess what P(E|G) means? Notice that the G and E symbols have been interchanged. P(E|G) represents the probability that the evidence E would have been produced if G were true. In other words, if it were the case that God exists is true, then P(E|G) is the a priori probability that the evidence E would have been produced. Similarly, if it were the case that God does not exist is true (i.e., G is false and G* is true), then P(E|G*) is the a priori probability that the evidence E would have been produced. Examples will help navigate through this menagerie of probabilities, but first, let's get to the much anticipated Bayes' theorem.

Wishing to protect you, the gentle reader, from a formal mathematical derivation of the theorem, I will take a more heuristic, intuitive approach to constructing Reverend Bayes' formula. (At this point I make the pledge that what follows is as mathematically challenging as this analysis is going to get. Once we have Bayes' theorem up and running, the road to follow will be somewhat less bumpy.) So let me begin by coaxing you back into the vibrant world of coin tossing. Imagine I own two special coins, each of them highly biased. That is, having tossed these coins many a time over the years, I have found that one of them usually (although not always) lands heads up, whereas the other usually (again, not always) lands tails up. We will refer to these as the heads-favoring coin and the tails-favoring coin, respectively. Now, I am holding one of these coins in my hand, but you don't know which one it is. I then ask you to assess the probability that I am in fact holding the heads-favoring coin. To help you in your task, I toss the coin just one time. It lands heads up.

So what is the probability that it's the heads-favoring coin? What would be your thinking on this? It seems intuitively reasonable to conclude that since the coin landed heads up, this evidence would point you more toward believing it to be the heads-favoring coin than the tails-favoring coin. But then you can't be certain that it's the heads-favoring coin since the other coin *does* on occasion land heads up.

Let's borrow the notation we have set up for our God exists proposition and apply it to this more mundane matter. So we will now define Propositions G and G* as

G = It is the heads-favoring coin that I am holding.

G* = It is not the heads-favoring coin that I am holding.

Equivalently, G* can be expressed as

G* = It is the tails-favoring coin that I am holding.

The evidence you have, denoted E, is the fact that on a single toss the mystery coin landed heads up. So

E = The coin on a single toss landed heads up.

I am asking you to come up with the probability that G is true given that you have the evidence E. That is, what is the numerical value of $P(G|E)$?

Let's begin by expressing the known bias of these coins in math-speak. If I am in fact holding the heads-favoring coin (i.e., if G is true), then the probability of a single toss resulting in heads is large. That is,

$P(E|G)$ is large.

On the other hand, if I am in fact holding the tails-favoring coin (i.e., if G* is true), then the probability of a single toss resulting in heads is small. That is,

$P(E|G^*)$ is small.

So here's our first intuitive step. Let's say that the larger P(E|G) is (i.e., the more biased the heads-favoring coin is known to be), the greater the probability that the mystery coin is in fact the heads-favoring coin given the evidence of a heads-up toss. That is, the value of P(G|E) increases as the value of P(E|G) increases. The simplest mathematical way of expressing this assumption is

$$P(G|E) \propto P(E|G)$$

where the symbol $\propto$ means "is proportional to." So the bigger P(E|G) is, the bigger P(G|E) gets. The analogous formula for Proposition G* would be

$$P(G^*|E) \propto P(E|G^*)$$

meaning the lower the chance that the tails-favoring coin would produce a heads-up result (i.e., the more biased it is toward tails), the lower the probability that the mystery coin is the tails-favoring coin given the evidence of a heads-up toss.

So far, so good. But how do the prior probabilities P(G) and P(G*) enter into the formula? You'll recall that these are the preevidence (pre–coin toss) probabilities attached to Propositions G and G*. Well, before I tossed the coin, did you have any preestablished notions about which of the two biased coins I was in fact holding? Perhaps someone had warned you, "If that guy does the dumb thing with the coins, remember that he's usually holding the tails-favoring coin—besides, I heard he might have spent the other one." Or perhaps you have no preestablished belief and you consider it a 50-50 situation. Whatever your preevidence probability for Proposition G might be, let's view it as a kind of head start for G in the race between G and G* to achieve the greater postevidence probability. That is, the greater the preevidence probability P(G) you come up with, the greater the postevidence probability P(G|E) will wind up being for the same evidence. So let's now revise the formulas to reflect this requirement. The simplest way to do this is as follows:

$$P(G|E) \propto P(G) \times P(E|G)$$

with a similar formula for G*:

$$P(G^*|E) \propto P(G^*) \times P(E|G^*).$$

Well, we're almost there, but an equation without an equal sign isn't much use. How do we get rid of the insubstantial "$\propto$" in favor of a good old "="? Well, here's how. We know that

$$P(G|E) + P(G^*|E) = 100\%$$

because the mystery coin has to be one or the other of the two biased coins; that is, there is a 100 percent probability that the coin is either the heads-favoring coin or the tails-favoring coin. So in converting our expression for P(G|E) into a solid equation, we need to insert a scaling factor that ensures the postevidence probabilities add up to 100 percent. Doing this, the equation we wind up with is

$$P(G|E) = \frac{P(G) \times P(E|G)}{[P(G) \times P(E|G) + P(G^*) \times P(E|G^*)]}$$

where the part in the square brackets, [. . .], is the scaling factor. Et voila—Bayes' theorem. This formula tells you how to take into account the evidence, denoted E, in calculating the posterior probability, denoted $P(G|E)$, starting from the preevidence prior probability, denoted $P(G)$. In other words, Bayes' theorem defines the way to adjust a probability based on the evidence. So if your prior probabilities corresponding to the two biased coins were simply 50 percent and 50 percent and the heads-favoring coin had historically produced heads up 90 percent of tosses, whereas the tails-favoring coin had produced heads up for only 5 percent of tosses, then

$P(G) = 50\%$

$P(G^*) = 50\%$

$P(E|G) = 90\%$

$P(E|G^*) = 5\%$

and, after the evidence of the single toss, you would apply Bayes' theorem to calculate that

$$P(G|E) = \frac{50\% \times 90\%}{(50\% \times 90\%) + (50\% \times 5\%)}.$$

That is, you would conclude that the probability I am in fact holding the heads-favoring coin is

$$P(G|E) = 94.7\%.$$

Allow me one more quick refinement to my expression of Bayes' theorem before we proceed. It will lighten our load later on in the analysis. Because P(G) and P(G*) must add up to 100 percent (for the same reason that the two postevidence probabilities needed to add to 100 percent), I can replace P(G*) in the formula with $100\% - P(G)$. So the version of Bayes' theorem with which we shall proceed is

$$P(G|E) = \frac{P(G) \times P(E|G)}{P(G) \times P(E|G) + [100\% - P(G)] \times P(E|G^*)}.$$

I hope you are well after all of this and have not regressed to a time when teachers considered foot rulers and pencil boxes to be weaponry. But you know, this morass of math is not as impenetrable as it may seem. Before returning to matters divine, let's work through another terrestrial application of Bayes' theorem—but this time no more of that tedious coin tossing.

> American counterintelligence agents believe that they may have captured the infamous Canadian master spy Le Mole. They are not certain that their detainee, Jean-Paul Poisson, is in fact Le Mole, but they think there is a relatively good possibility that he is their man. They decide to give

him a lie detector test in the hope of increasing their confidence that the master spy has been captured. During this test, Poisson denies that he is Le Mole. "Le Mole, moi? Mais non." The lie detector test indicates that Poisson is lying.

How would the agents use Bayes' theorem to determine how the results of the lie detector test should influence their assessment of the probability that Poisson *is* Le Mole. Let's cast this situation in the mathematical language we have just established. These are the steps:

1. Let G now represent the proposition that Poisson is Le Mole.
2. Since G* is by definition the negation of G, then the Proposition G* is that Poisson is not Le Mole.
3. E is the new evidence provided by the lie detector test. That is, E = Poisson denies that he is Le Mole, and the lie detector test indicates that he is lying.
4. Let us assume that the agents are aware that lie detector tests are not completely reliable. The probability of a test giving a false positive (i.e., indicating that the subject is lying when he is actually telling the truth) is, say, 5 percent. If Poisson were telling the truth, then G* would be true and the test result, E, would have been a false positive, so

$$P(E|G^*) = 5\%.$$

5. Now, if G is true (i.e., Poisson is Le Mole), then what is the probability that the lie detector test result, E, would have been produced (i.e., that the test produced a true positive)? This probability is denoted P(E|G). Let's assume it is known that the probability of a false negative result for this lie detector test is 15 percent (i.e., this is the probability that the test fails to detect a lie). This means that the balance of probability (i.e., 100 percent less 15 percent, or 85 percent) is the probability that the test gives a true positive. That is,

   $P(E|G) = 85\%.$

6. Let's assume that prior to the test, consideration of circumstantial evidence had made the agents 75 percent confident that Poisson is Le Mole. That is, the prior probabilities are distributed such that there is 75 percent probability that G is true, so

   $P(G) = 75\%.$

Finally, what the agents wish to determine is the probability that G is true, that is, that Poisson is their man, now that the lie detector test results are in. This posterior probability is denoted P(G|E). Now it is just a matter of taking Bayes' theorem as presented and replacing each symbol with the corresponding probability numbers as follows:

$$P(G|E) = \frac{75\% \times 85\%}{[75\% \times 85\%] + [(100\% - 75\%) \times 5\%]}.$$

Whipping out the calculator, the result is

$$P(G|E) = 98\%.$$

So the lie detector test has had the effect of increasing the probability that the agents have the right man from 75 percent to 98 percent.

Ah, you may say, this seems like a very systematic and mathematical way of deriving a posterior probability that takes into account the evidence, but where did that prior probability number of 75 percent, that is, P(G), come from? You have a way of getting to the end point of a posterior probability only if you have the luxury of an appropriate starting point in the form of a prior, preevidence probability, and where does *that* come from? You may have noticed that I glossed over that part of the analysis by saying that circumstantial evidence had led the agents to a prior probability of 75 percent that Poisson was Le Mole. Where's the clever mathematics that produced *this* number from the aforementioned circumstantial evidence? If this indeed was your observation, then you have identified the most commonly cited weakness of the Bayesian approach to generating probabilities: It demands the availability of a prior probability, and only once that the prior probability is in hand does it then allow the use of evidence to update that probability. The frequentists' taunt is, "Where did the prior probability come from, Mr. clever Bayesian person?"

The frequentists do have a point. A preevidence probability must be estimated in the Bayesian approach. In an ideal Bayesian world, we would wish to develop a preevidence probability that reflects complete ignorance about the issue at hand and then use Bayes' theorem to update that probability in light of all the available evidence. But how is complete ignorance represented by probabilities? While some ignorance—that is, being in a situation where uncertainties are present—provides the very need for working with probabilities, what do we mean by complete ignorance? One solution is to say that complete ignorance is the situation in which you attach an equal probability to each possible outcome. For example, if there are two possible outcomes, only one of which can be true (but you don't know which one), then complete ignorance about the outcome could be represented by attaching a 50 percent probability to each of them.

This is an appealing approach, but it has its flaws, as the following example shows. Consider an irregular stone with eight flat surfaces, each with a number painted on it: 1 through 8. Now assume that on six of the eight faces, the number is painted in blue, while on the remaining two faces, the number is painted in red. The stone will be thrown, and the outcome of interest to me is whether it will land with a blue number down to the ground or a red number down.

(We presume to know that one flat face of the rock will land flush to the ground.) How would I represent complete ignorance about this future outcome, red or blue number down? Based on the simple approach just articulated, I would represent complete ignorance by attaching a 50 percent probability to each of these two possible outcomes: blue or red number down. However, someone else may interpret this simple rule of complete ignorance differently. She may identify the outcomes of the stone throw as the events that the number facing down is 1, 2, 3, 4, 5, 6, 7, or 8. For her, complete ignorance would be the attachment of a 12.5 percent probability to each face of the stone, which is equivalent to attaching a 75 percent probability to the outcome of a blue number down (recalling that six of the eight numbers are painted in blue) and a 25 percent probability to a red number down. That is, she deviates from the 50-50 assessment.

She and I would then disagree about what constitutes complete ignorance and would engage in an argument about which of us is the more ignorant. (Being a gentleman, I would of course acquiesce ultimately and concede her greater ignorance.) My point is that equal apportionment of probabilities to all possible outcomes can be an ambiguous proposition when promulgated as a general rule for representing complete ignorance.

Bayesian analysts would generally agree that the choice of prior probabilities must be considered and justified on a case-by-case basis. In this spirit, let's leave behind the irregular stone and the case of Le Mole to return to

Proposition G = God exists.

Here, I think that the expression of complete ignorance is a good case for the 50-50 argument. That is, to profess the maximum ignorance on this issue is to assess the probability that G is true at 50 percent. Thus the balance of probability, the other 50 percent, is associated with the proposition that G* is true, where

Proposition G* = God does not exist.

So

$P(G) = 50\%$.

This is the perfect, unbiased expression of agnosticism. Now, I suppose a troublemaker might counter that we should really lay out a spectrum of optional beliefs about the nature of God, one for each alternative religious view, and then attach an equal probability to each while reserving an equal, residual probability for the proposition that God does not exist. So we might begin with, say, nine versions of God corresponding to various religious perspectives, to each of which we attach a 10 percent probability, with the 10 percent balance attaching to the proposition that God does not exist. This is not the way to go, in my view, for two reasons. First, the number of alternative sets of beliefs about the specific nature and preferences of God is vast, reflecting alternative faiths, sects within faiths, and people within sects. Thus one could easily depress the prior probability of the no-God proposition to an arbitrarily small value by simply invoking greater numbers of God options. In this way, the single religious perspective of the world's atheists would be probabilistically swamped through application of an arbitrary, mathematical process. For instance, if we came up with 999 God options, the prior probability of the no-God proposition would become a mere 0.1 percent, or 1 in a 1,000. For that matter, any single religious view would be swamped also.

Second, the question of the specifics of the person-God's nature is secondary in my mind to the question of his very existence. As noted in chapter 2, the monotheistic faiths surely have more commonalities than differences in their perception of God. Thus I will stand by Propositions G and G* as the optional truths under analysis, to each of which a 50 percent prior probability will be attached. Indeed, from the perspective of prior probabilities, I find the God issue intuitively easier and less ambiguous to deal with than the irregular rock issue. Such is the strangeness of the Bayesian world.

For the mathematically adventurous reader, D. V. Lindley's *Introduction to Probability and Statistics from a Bayesian Viewpoint (Part 1: Probability)* (1965) is a solid introduction to the Bayesian perspective on probability theory.

Having established definitions and formalism, we may now proceed.

Chapter Conclusions:

Probability, in the Bayesian outlook, is a measure of partial belief or level of confidence in the truth of a proposition.

Probabilities are applied where uncertainties exist.

A probability is a number in the range of 0 percent to 100 percent.

Bayes' theorem provides a systematic means of modifying probabilities based on consideration of evidence.

Maximum a priori (preevidence) ignorance about whether or not God exists can be represented by attaching a 50 percent truth probability to Proposition G.

five

The Bayes Craze

Questions for This Chapter:

What would be some practical applications of Bayesian inference?

Is it technically legitimate to represent degrees of human belief in mathematical form?

If the previous chapter was your first introduction to Bayesian thinking, then I suspect you have formed a few questions—perhaps even some doubts. After all, is it realistic that the awesome machinery of probabilistic mathematics could be used to power a concept so fluffy and blond as degrees of belief? In the Bayesian world, this is precisely what a probability represents: a degree of belief or level of confidence that some proposition is true.

If the popularity and breadth of application of a mathematical method is any measure of its validity, then Bayesian methods should be assessed very positively. The 1980s saw the beginnings of an explosion in practical applications of Bayesian inference, that is, the use of Bayes' theorem to assess uncertainties in decision making. To give a flavor for the breadth of Bayesian applications, we have

- computer-based systems for medical diagnosis
- bases for legal inference and judicial decision making
- risk analysis of industrial facilities
- design of robotic vision and sound detection systems
- archaeological dating of prehistoric monolithic structures
- PC software to facilitate personal time and communications management

to mention but a few.

Of course, popularity is no measure of technical or scientific validity. To quote the physicist Luis Alvarez, "There is no democracy in science." You cannot, for example, vote for a particular scientific or mathematical theory from a point of utter ignorance the way you can vote for, say, a presidential candidate. So rather than relying smugly on the fact of the broad application of Bayesian methods as justification for our use of them, let's consider their pros and cons a little further—and *then* proceed smugly.

It must be conceded, the idea that one can mathematically represent the workings of the human mind has a certain arrogance to it. Understanding of the human mind is generally considered to be the territory of psychologists and psychiatrists, and neither of these professions is known for its prowess with the slide rule. As a general principle, the more complex a system, the greater the difficulty in using mathematics to predict its behavior. For example, while the mathematics of quantum mechanics can do a very precise job of modeling the behavior of the simplest atom—the hydrogen atom, consisting of one proton and one electron—attempts to solve the mathematics of the next-to-simplest atom (helium) require us to resort to various approximate solutions since exact mathematical answers to the quantum mechanical equations cannot be obtained. Much of physics and all of engineering require these sorts of approximations to be made because the underlying systems are so complex. So if even a helium atom begins to present mathematical difficulties, how realistic is it to mathematically represent degrees of belief, which are products of the human thought process and all its incomparable complexities? Surely simple little equations such as Bayes' theorem and the other basic formulas of probability theory cannot hope to capture the intricacy of human thought.

It must be conceded, the idea that one can mathematically represent the workings of the human mind has a certain arrogance to it.

So is the very attempt to represent degrees of belief with probability theory an absurd one? While it's all well and good to rate your colleague's girlfriend on a scale of 1 to 10, surely you are not so presumptuous to consider that numerical quantity a mathematical object, subject to some set of strict, well-established algebraic theorems. Unlike the girlfriend rating system, the degrees of belief of Bayesian inference *are* identified with well-established mathematical objects called probabilities.

There is certainly something to this criticism, and it's worth exploring. First, I'll mention that probability theory is viewed by Bayesians as a *prescriptive* basis, and not a *descriptive* basis, for quantifying degrees of belief. That is, probability theory is considered to provide a rational basis for updating degrees of belief in light of evidence but does not claim to necessarily emulate the way a real person would update her beliefs based on the same evidence. For example, had the American counterintelligence agents not applied Bayes' theorem but, rather, simply applied their intuition in revising the probability that Poisson was Le Mole after the lie detector test, they would not necessarily have arrived at the same postevidence probability. Put another way, Bayesian methods prescribe how degrees of belief *should* be updated and not necessarily how they *would* be updated absent the math. Anyone claiming that they have a mathematical system capable of synthesizing the human thought process would indeed be bold. Bayesians are not that bold.

Characterizing Bayesian inference as a prescriptive method takes the heat off a little, but it's still pretty hot here under the sunlit magnifying glass of the traditional frequentists. Even if we accept Bayesian inference as a prescriptive method only, one can still question the very notion that degrees of belief can be hardwired in some way to objective evidence.

Consider two individuals. Let's assume that they are not mad, inebriated, or otherwise irrational. If these individuals are presented with precisely the same evidence and other background information, will they necessarily agree on the numerical probability that should be attached to some specified proposition? Let's exclude from consideration topics that are purely matters of taste. As a nonprobabilistic example, I may give a score of 3 to my colleague's girlfriend, whereas he might give her a 9. This is simply a matter of taste, and we would not expect the scores to coincide. However, if my colleague and I were asked about the probability that the neutrino particle is massless, would we produce the same probability number? Assume that we are both aware of the same research, models, predictions, and data. Would this be sufficient to ensure that we agree on the value of the implied probability?

Some argue resolutely that any two rational people confronted with the same evidence would produce the same probability.

This is a very tough question, and there is no agreement in the Bayesian community about the answer. Some argue resolutely that any two rational people confronted with the same evidence would produce the same probability. Others think this a pipe dream, even as a hypothetical, idealized situation. Ask two individuals the probability that their opposing soccer teams will win next Saturday, even when they are in possession of the same facts, and I doubt that you would get agreement on the answer. Ah, but this is pure bias, you may say. In this example, the facts they *do* have are trumped by their typical psychotic, violent, all-consuming desire for their soccer team to win. As someone who hails from the city named for Manchester United Football Club, I see this response as fair comment. Yet I wished to present an extreme example of bias. In other situations, bias can be far more subtle and often inextricable or even indistinguishable from the evidence.

My own experience leads me to argue against the notion of equal evidence leading to equal probabilities. In the engineering risk analysis field, I have been involved in many exercises to elicit from experts the truth probabilities of various technical propositions. These experts were typically nuclear, chemical, civil, mechanical, and electrical engineers; human reliability analysts; seismic experts; meteorologists; and chemists—and not always soccer fans. Often did I encounter situations in which two apparent experts, both familiar with the same relevant evidence, would be at the point of fisticuffs over the assignment of appropriate probabilities. Each of them had the correct viewpoint, and each of them knew it! Happily, there are mathematical means of aggregating disparate probabilities from a pool of experts, but it is nevertheless a disconcerting experience to witness the bloodiness of probabilistic dispute.

To a limited degree, the type of bias associated with the soccer fans *can* account for the differences in opinion between technical experts, but it would be glib to blame every dispute on bias. Throughout our lives, we humans individually accrue an enormous databank of experiences and information. To presume to inform someone which evidence is strictly relevant to a given proposition is generally unrealistic. While a judge may take such liberties in his court when the proposition is a legal one, such cheekiness would be inappropriate in most other settings. We humans continually make cross-inferences between our fields of knowledge, and the line between bias and legitimate evidence is a dull, fuzzy one. Another way of looking at it is that the hypothetical situation will never occur in which two people have access to exactly the same evidence. Life thrusts too much information at us. It is therefore a vacuous exercise in my view to speculate on whether access to the same evidence should always lead two people to produce the same probabilities.

Have I just admitted that Bayesian probabilities are utterly subjective and so worthless as aids to . . . to whatever is in need of aid? Of course not—why would you think that?

I just wish to appear unbiased before I proceed to extol the merits of Bayesian inference.

It is certainly the case that Bayesian probabilities have a subjective element. A degree of belief is a subjective notion. This is not to say that there are no constraints on the creation of probabilities: The rigid form of Bayes' theorem itself injects some order into the process. In a sense, partial subjectivity is an inevitable attribute of probabilities. After all, recall that the value of probability theory lies in its ability to aid the making of decisions under uncertainty, and uncertainty is surely an attribute of the perceiver; that is, uncertainty must have a strong subjective element.

So how does this subjectivity compromise the defensibility of probabilities? Well, to answer this question you must consider the alternatives. If we were presented with the choice of a completely objective, evidence-driven means of assessing uncertainty versus the partially subjective approach of Bayesian inference, I think most of us would choose the former. Disappointingly, these are not the options. The *real* options are as follows:

Option 1: Try to deal with uncertainties in an intuitive way, balancing the facts in your mind and relying on an ingenious facility for being right to reach your decision.

Option 2: If you're a frequentist and the problem at hand cannot be understood in terms of coin tossing and other stochastic (random), statistical processes, completely replace the problem with one you know how to cope with. Then publish the results and extol the virtues of traditional frequentist statistics.

Option 3: Define your uncertainties in terms of propositions and associated probabilities. Try to minimize subjectivity through justification of assumptions and probabilities, as well as through systematic application of Bayes' theorem where practical. In this way, your assumptions and reasoning are transparent and scrutable so that anyone who wishes to take you to task can challenge specifics of your analysis. They could even reproduce your analysis with their own subjective assumptions, replacing yours to see how the conclusions are affected.

Now you choose. To complete this chapter, I'd like to present something I call "The Convenient Means of Summarizing This Chapter's Conclusions Vignette."

THE CONVENIENT MEANS OF SUMMARIZING THIS CHAPTER'S CONCLUSIONS VIGNETTE

Scene: Two individuals in the stands of an enormous, otherwise empty professional sports stadium.

Characters: The Author (as himself) and Frankie the Frequentist.

FRANKIE: You ignorant, ignorant fool.

AUTHOR: I'm sorry, what?

FRANKIE: Oh . . . I was just warming up. Have we started yet?

AUTHOR: Eh, let's start now.

FRANKIE: Okay. [*Smirking.*] So I understand that you've set yourself the task of calculating the probability that God exists.

AUTHOR: [*Beaming.*] Yes, yes I have.

FRANKIE: I think you're crazy. Besides, didn't some fellow named Richard Swinburne already consider the probability of God?

AUTHOR: Oh, he dabbled on the periphery of probabilities in an abstract, scholarly sort of way but never got the dirt of real numbers under his fingernails as far as I know.

FRANKIE: Then there's the intelligent design theory people who . . .

AUTHOR: Yes, yes, but they look at cosmology and biology as the basis for their God considerations. I'm not going to look for God in scientific gaps. Besides, *any* scientific theory can be juxtaposed by an alternative explanation involving the actions of a supreme intelligence. Thus we can assert that physical phenomenon X—insert your phenomenon here—occurs because God so wills it. For example, who needs a theory of gravity? God *wills* things to fall . . . that's the only theory you need! Newton . . . Shmuton. No, no, that's not the path I plan to take. I want to do a pragmatic analysis of the things that are really germane to a religious outlook on life. Didn't you read chapter 3?

FRANKIE: I didn't know I had to prepare. Anyway, it's all this Bayesian nonsense that I have the greatest problem with. You've openly admitted that probabilities as viewed by Bayesians are simply degrees of belief: levels of personal confidence in the truth of a proposition. What complete fluff! Where's the objectivity in that? Why even bother with Bayesian inference if you and some other Bayesian could be presented with identical evidence and yet end up with different results?

AUTHOR: Well, as a frequentist, how would *you* handle the situation?

FRANKIE: Which situation?

AUTHOR: Determining the probability that God exists?

FRANKIE: There's no such thing as the probability that God exists, so how can I estimate it? Probabilities come from repeated trials of events, like the coin tossing you've been deriving boundless fun from. Proposition G has nothing to do with repeated trials. You know, sometimes the appropriate conclusion is that there *is* no meaningful question to answer, and so trying to answer it is a vacuous exercise.

AUTHOR: So you just throw up your hands?

FRANKIE: Yes. That can be the knowledgeable resolution.

AUTHOR: I'm glad you're not my physician. Say you were a doctor who was uncertain about a diagnosis. Wouldn't probabilistic thinking play any part in what remedies you'd try first?

FRANKIE: Well, if by probabilistic thinking you mean my beliefs under uncertainty, then of course it would.

AUTHOR: So you wouldn't just throw up your hands if you were in that position. You'd need to make a decision as best you could, and your varying degrees of belief about the alternative prospective diagnoses would be critical considerations: Which diagnosis is more likely than another?

FRANKIE: Yes, you do have a powerful grasp of the blatant.

AUTHOR: Say you had to make a decision that involves many factors and uncertainties. If you had a means of systematizing your thought process, wouldn't you use it?

FRANKIE: Yes, I would.

AUTHOR: Well, that's the only claim I'm making for Bayesian methods: They provide a systematic framework for considering uncertainties.

FRANKIE: But you Bayesians bandy about your probabilities as if they're immutable, universal constants like the speed of light or the value of pi, dictated objectively by hard evidence.

AUTHOR: Some are guilty of that, but not me. All I want to do is make my reasoning as open and transparent as possible to anyone who wishes to follow it and assess its rational probabilistic implications.

FRANKIE: Maybe you're not so naive after all. [*Aside.*] This script is shameless.

AUTHOR: Thanks, Frankie. Ultimately, I'll produce a probability that Proposition G is true. You may not like some of the assumptions or conclusions I make along the way, and you may even be engaged enough to adjust those assumptions for yourself to see how the final numbers are affected. You'll be able to do that because the probabilistic thought process will be plain and scrutable. That's the real beauty of Bayesian analysis. That's not to say I won't try to be as objective as possible and defend my assumptions. But mainly I just want you to know how I get to the results.

FRANKIE: Okay, I'll go along with that for now. By the way, how will you use the truth probability of Proposition G once you have it?

AUTHOR: I have some ideas about that, but you'll have to wait and see.

FRANKIE: The anticipation is unbearable. And why are we sitting in an empty sports stadium?

AUTHOR: [*With undue gravitas.*] It's a metaphor for the grandiosity of our dialogue.

FRANKIE: Good grief.

Curtain falls.

Chapter Conclusions:

Bayesian inference has proven a powerful analysis tool in a range of technical disciplines.

Bayesian inference is a pragmatic tool for ordering one's thoughts on complex issues and facilitating decision making under conditions of uncertainty.

---

six

Down to Business

Question for This Chapter:

What are the evidentiary areas that we will consider in assessing the truth probability of Proposition G?

Having emerged with your diploma in Bayesian Inference 101, you are ready to put it to use. First, recapitulation of two key points will be helpful. I will use the voice of first-person plural because (a) I am confident that you have gone along with my reasoning so far, and (b) it gives the impression of academic rigor. These are the key points:

1. We have determined that a probability is a metric of degree of belief, and as such it inevitably reflects a

degree of subjectivity. However, probability theory, and in particular Bayes' theorem, provides a basis for systematization of the thought process in addressing uncertainties. The result is that the scrutability, transparency, reproducibility, and defensibility of that thought process is maximized. (See chapters 4 and 5.)

2. We have established the irrelevance to our analysis of issues related to the general cosmological, physical, and biological nature of our universe. (See chapter 3.)

This is where we stand and from where we may now proceed. On point 2, you may think that discounting the relevance of the physical universe has the effect of somewhat narrowing the field of prospective evidence germane to Proposition G. This is true. Nevertheless, in this chapter I will attempt to identify those areas of evidence that I *do* consider to be prospectively relevant. In chapter 3, I introduced a television analogy. I suggested that looking for evidence of God in the laws and general phenomena of nature is like breaking open a television set to look for the tiny actors inside. Cathode ray tubes and TV shows are completely disparate but fully compatible descriptions of the phenomenon of television. If the *cathode ray tube perspective* represents the physical and biological description of the natural world, then what I seek to do in this chapter is to identify the *TV shows perspective.*

Let's begin by returning to the notion of the person-God discussed in chapter 2. This is the God of the major monotheistic faiths. Not all people of faith agree on the specific characteristics and preferences of the person-God, but beliefs tend to coincide in important aspects. The person-God is compassionate, fair, and merciful. He is concerned with each and every human life and sustains us individually. We can communicate with him personally, through prayer or perhaps through deed. He communicates back with us through our life experiences, through our conceptions of what is moral and good, and perhaps even via more direct conduits such as revelations or other types of spiritual experiences. These are the attributes of the person-God around which our analysis revolves. Given this understanding of what we mean by "God," the remainder of this chapter is dedicated to identification of a series of evidentiary areas that I think bear on Proposition G. In each area I will give a brief outline of what I consider to be the key issues and alternative perspectives. Of course, each of these areas deserves to be, and has often been, the subject of voluminous analysis and discourse. Nevertheless, I will be ruthless in adhering to my stated objective of conducting a quick, pragmatic, nonscholarly, get-on-with-it sort of analysis. So this chapter will introduce the evidentiary areas, then in the next chapter we'll subject them to the rigors of our daunting Bayesian arsenal.

EVIDENTIARY AREA: THE RECOGNITION OF GOODNESS

We humans seem to have a sense of what is good and what is evil. We certainly encounter areas of ambivalent gray in our lives, but even the recognition of grayness requires an understanding that it is a mixture of black and white components, a mixture of good and evil. Helping a hurt child is good, while torture is evil. Goodness can have various manifestations and means of implementation: kindness, mercy, compassion, duty, valor, and honesty, to name some. Yet the dispassionate, indifferent world of physical systems would seem to have no place or language for goodness. Therefore, *goodness* is not generally encountered as a learning unit under Electromagnetism 101. Theists would argue that the existence of goodness is clear evidence of God since from where else could the innate recognition of the absolute of goodness originate? Recall Genesis 1:27, "So God created man in his own image," and so innate in us is a recognition of the good. The notion of what is good, notwithstanding some cultural variation, appears to largely transcend societies and eras, and so cultural, relativistic explanations would be inadequate to explain the universal recognition of goodness. Hurting a child is a universal evil. This is the so-called moral argument for God's existence.

As would be expected, the theists are not given a free ride here since there are scientific and philosophical descriptions of goodness that do not rely on the existence of a person-God. For example, Darwinians note that acts of apparent altruism are not exclusively human traits. The field of ethology, which compares human and animal behavior, has suggested natural selection mechanisms for the emergence of emotions and values in humans. One argument goes that mammals, who produce very few eggs compared to other animal species, have good pragmatic reason to be very protective of their young. This then passes for compassion. On a larger scale, it is to the advantage of every mammal to perform acts that protect its own clan from danger, and so the practical value of this pattern of behavior has resulted in what we consider to be human altruistic values.

The dispassionate, indifferent world of physical systems would seem to have no place or language for goodness. Therefore, goodness is not generally encountered as a learning unit under Electromagnetism 101.

Certain philosophers have rejected this notion that goodness derives from nature, and yet curiously, they have done so without recourse to reliance on God. Most notably, the early 20th-century British philosopher G. E. Moore considered goodness to be a "simple, undefinable, nonnatural property." I think there was a beauty to the 20th-century British and American schools of philosophy. People such as Moore, Russell, Wittgenstein, and Ryle took an absolutely no-nonsense approach to the philosophical matter of "What is the reality?" and focused purely on the importance of language. Instead of contemplating the true inner nature of the world as philosophers had done for 25 centuries, they seemed to throw up their hands at such profundity and instead focused on questions such as "Did what you just say have any meaning, and if so, what is it?" You can imagine how effectively that sort of philosophy can shut down a conversation and indeed vaporize advertising and politics with brutal efficiency. Thus was the external, objective world removed from the spotlight of philosophy and replaced by our means of *describing* that world.

But back to Moore. When he asserted that goodness is nonnatural, he was not invoking the supernatural. He meant that goodness could not be understood in terms of human desires, such as comfort, security, and pleasure, nor could it be broken down into component parts. For example, if we did reject his idea that goodness cannot be understood in terms of elemental components and causes and defiantly proposed that "pleasure is a form of good," then Moore might ask, "But is pleasure itself good?" Well, we might respond that this is a legitimate question to ask. "Ah," he would then say, "but if that question makes sense, then I have demonstrated that pleasure and goodness are not the same thing." (When I suggested that his school of philosophy was a no-nonsense one, I meant, of course, by philosophers' standards.)

Moore attached the term *naturalistic fallacy* to any belief that goodness can be defined in terms of natural properties such as pleasure and comfort. To ensure that the term *naturalistic fallacy* should not be construed to support supernatural explanations of goodness, some have suggested that it be retitled the *definist fallacy.* This rubric would then capture both the naturalistic fallacy and its complementary *metaphysical fallacy,* the belief that goodness can be identified with metaphysical notions such as God. While Moore's conclusions would preclude us from understanding goodness in terms of evolutionary biology, they do not appear to answer the question of why goodness is universally recognized. Moore acknowledged this problem and suggested the existence of something called *moral intuition,* a standard human attribute like, say, optical vision or a sense of smell.

Leading to Moore's ideas had been a long history of philosophical discourse on the nature of goodness. Plato's dialogues have Socrates arguing for the rationality of virtuous behavior. Aristotle believed that people are truly virtuous only if they enjoy performing acts of goodness. Kant in the 18th century opposed Aristotle's position by concluding that virtue should be attributed only to those who do good out of duty despite the pain of well-doing. These philosophers saw goodness as an absolute but not necessarily deriving from a person-God.

In contrast to the view that goodness is absolute, there is also the relativist perspective. This variety of relativity has nothing to do with Einstein and refers to the notion that what you consider good and moral depends on your perspective and experiences. This position is exemplified by the contemporary American psychologist Jonathan Haidt in his paper "The Emotional Dog and Its Rational Tail: A Social Intuitionist Approach to Moral Judgment." He argues that we all acquire our moral intuition largely from interaction with peers, especially in our teen years. He points out that certain moral values can be so widespread in a community that they give the impression of being objective truth. This, he asserts, accounts for a history of morally different minorities (such as communists, conservatives, Jews, and homosexuals) being purged, expelled, or even killed by cohesive moral communities. Sadly, God's will is sometimes, although not always, the excuse for such damaging moral cohesion. This relativist view of goodness is in stark contrast to theological and historic philosophical perspectives.

So while the recognition of goodness is not a home run for Proposition G, it is an important area of evidence for us to consider in our analysis.

EVIDENTIARY AREA: THE EXISTENCE OF EVIL

In my experience, the single most troubling issue for people of faith is the fact that evil exists, and exists in abundance. Setting aside the epistemological issues of goodness discussed as the previous evidentiary area, the down-to-earth fact of the matter is that there exist horrible evils in the world. Day by day, minute by minute, people are suffering through war, torture, disease, famine, neglect, and other evils. How can this state of the world be reconciled with the religiously ascribed attributes of a person-God who is

- omniscient—therefore, he is fully aware of the evils in the world
- good—therefore, he is fair and merciful
- omnipotent—therefore, he can effect any change he wishes

Does the presence of evil in the world require us to settle for two out of three of these attributes since all three in combination would surely preclude the existence of evil? How are we to interpret Genesis 1:31, "And God saw every thing that he made, and behold, it was very good"? "Very good" seems to be an exaggeration. "Largely good," perhaps; "bits of it were good," more realistically—but "very good"? I suppose we must assume that God was temporarily reining in his omniscience to view just a brief moment in time in the world's prehistory.

Although God's allowance of evil and the suffering it causes may be bewildering to most people of faith, and may be the trump card for atheists, there are those faithful who have conceived rationales for evil. The contemporary British theologian and philosopher Richard Swinburne argues in his book *Is There a God?* (1996) for the "free will defense" of evil. First, he categorizes evils into two sets: "moral evil," the source of which is the immoral actions of people; and "natural evil," a category encompassing natural hazards such as disease, destructive weather, earthquakes, and the like. I would add to this latter category the hazards associated with man-made systems, such as aircraft crashes, car accidents, industrial accidents, and other random phenomena brought about by well-intentioned human error or even by random failures of equipment. All these evils can result in enormous human suffering.

In the case of moral evil, Swinburne argues that God has given us free will and genuine responsibility for the well-being of others. Thus to close off our option of evil is to compromise that free will and so trivialize our ability to opt for responsible choices. He expands this theory to capture natural evils. He does this by noting that natural catastrophes give us especially meaningful opportunities to make good choices, such as comforting the suffering or taking a stoic attitude to one's own misfortunes. Since the free will rationalization of natural evil and the free will rationalization of moral evil strain my own credulity to differing degrees, I will treat them as separate evidentiary areas for the purposes of the forthcoming Bayesian analysis in chapter 7.

Another perspective was offered by the late-20th-century Christian journalist Malcolm Muggeridge, who viewed evil as something that transforms the dull trek of life into something more akin to an exciting roller coaster ride or that injects drama into an otherwise tedious play. In his book *A Twentieth Century Testimony* (1978) looking back on his life he noted that "everything I have learned in my 75 years in this world, everything that has truly enhanced and enlightened my existence, has been through affliction and not through happiness." I suspect that this is a legitimate perspective for those of us who have not suffered excessively in life, but it is a viewpoint that I would be reluctant to share with, say, a Holocaust survivor.

A theological perspective on the existence of evil advocated by contemporary philosophers such as William Alston is that we humans lack the cognitive ability to understand God's moral reason for allowing evil. In other words, we're not smart or knowledgeable enough to comprehend the reconciliation of evil with the existence of an omniscient, good, omnipotent God. So as we view and seek to understand an evil world, we are little more than Arthur C. Clarke's apes taking ignorant jabs and sniffs at the monolith. Is there a grander logic of which we cannot be aware? Turning from the space odyssey of 2001 to the reality of 2001, it is an unsettling thought that the destruction of the World Trade Center served some greater good. Perhaps one of the slain would otherwise have spawned an evil son who would one day have brought about even greater suffering. If so, was there no better way of preventing the birth of this evil son, something more narrowly targeted? But then the question "Who are *we* to think it through?" is the crux of this perspective on evil. These are important evidentiary areas to consider.

EVIDENTIARY AREA: MIRACLES

Miraculous events play an important role in most versions of the Judeo-Christian faiths. The word *miracle* is used in various senses. It stems from the Latin *miraculum,* from *mirari,* which means "to wonder." In its most general sense, a miracle is any act performed by God. The *New Advent Catholic Encyclopedia* suggests one definition of miracles as "wonders performed by supernatural power as signs of some special mission or gift explicitly ascribed to God." In conformance with this general sense of the word, we can consider two classes of miracles:

1. Phenomena that could be construed as being brought about by natural forces but were in fact initiated or facilitated in some way by God. An example might be remission from a serious disease due to an act of God.

2. Phenomena that violate natural laws and indicate a blatant suspension of the physical workings of the universe. Raising of the dead would be an example.

The word *miracle* is commonly used in the narrower sense of class 2 only. These two types of phenomena are sufficiently distinctive to warrant separate analysis. Let's refer to them as *intra-natural miracles* and *extra-natural miracles,* respectively.

Intra-Natural Miracles

The person-God of the monotheistic faiths is presumed to influence worldly events in a very concrete manner. He answers prayers and can also be more proactive in manifesting his compassion for us. However, these acts may appear to be effected purely within the constraints of natural law. For example, our prayers may be answered for a friend's recovery from illness or perhaps for more trivial gains, such as a business success. These outcomes could be understood as natural phenomena.

The 20th-century British author Aldous Huxley considered miraculousness to be measured by the emotional reaction of the witness rather than by some property intrinsic to the observed phenomenon itself. So if my friend recovers from illness, then it is a miracle if I perceive it in wonder as a response to my prayer. This might be viewed as an unsatisfying and trivializing perspective on what constitutes a miracle. If indeed the miraculousness of an event is an attribute of the witness rather than the event itself, then the question of whether intra-natural miracles occur is trivially answered "yes." If they are believed to occur, then they *do* occur. While this approach is helpful in producing a definitive answer, the price paid is that the answered question was not worth asking. Therefore, we will *not* adopt this subjectivist definition of a miracle for the purposes of our analysis.

So if my friend recovers from illness, then it is a miracle if I perceive it in wonder as a response to my prayer. This might be viewed as an unsatisfying and trivializing perspective on what constitutes a miracle.

Now, if God is effecting miracles, then he is setting world events on a course that they would not otherwise have taken. Yet this tinkering principle is precisely the one I used in chapter 3 to dismiss the idea that God would override the natural laws he had created. If we dismiss the idea of God's ongoing interference with natural cosmological and biological systems on the grand scale, why not also dismiss the notion of microinterference with the physical world as it affects the lives of individuals? If he set the course, whether it be on the large scale or the individual scale, was it so imperfect as to demand continual adjustments? There are those inclined to produce logical analyses supporting God as a tinkerer. Perhaps it can be argued that each prayer is anticipated by God, and so his response was always part of the master plan. Perhaps the free will argument is relevant here: God intervenes on occasion to help us cope with some of the consequences of our decisions. This sort of reasoning can be very arbitrary. Suffice to say for now that God must surely influence events in our personal lives if he is the person-God of religious belief. The evidence of such influence is therefore legitimate for us to consider in addressing the truth of Proposition G.

Extra-Natural Miracles

Whereas some acts of God are apparently effected within the constraints of natural laws, extra-natural miracles involve the temporary adjustment or suspension of those laws. Many philosophers, such as Spinoza (see chapter 2), Immanuel Kant, and David Hume, have discounted the notion of phenomena external to nature and attributed accounts of such to flawed perceptions or analysis. Since to Spinoza, God and nature were indistinguishable, his position on this matter is no surprise, as a miracle would in a sense be self-violation on God's part.

Even among theologians, there is not uniform acceptance of the possibility of extra-natural phenomena. Some of them adhere to a theory of interpretation, in which it is the spiritual response to natural events that is the essence of a miracle. Again, this transfers the attribute of miraculousness from the event itself to the witness's perception of the event. There are others who believe that miracles are objective extra-natural phenomena but that they ceased to occur after the time of the Apostles, so all subsequent miracles are fakes or errors in interpretation. Preservation of the biblical miracles and rejection of modern-day miracles does have some pragmatic value, allowing certain important scriptural events to remain intact but precluding the need to take contemporary claims seriously. Then there are those who believe in extra-natural miracles past and present. This will be one of our evidentiary areas.

EVIDENTIARY AREA: RELIGIOUS EXPERIENCES

Have you ever looked at the vastness of the night sky or at a multihued sunset and felt a sense of awe? If so, was there an element to your experience that transcended natural wonderment? In the early 20th century, the American philosopher and psychologist William James produced a landmark work on the topic of religious experiences entitled *The Varieties of Religious Experience* (1994). James rejected many of his contemporaries' explanations of religious experiences, which at that time were often ascribed to hallucinations induced by a range of medical conditions from mere indigestion to nervous disorders. He attached spiritual significance to such experiences and described them in terms such as *feeling one's standing in relationship to the divine.* The word *experience* in this context is not to be understood in terms of experiencing a lower backache or 25 years in the garments industry. The word is used by James and subsequent religious philosophers to mean an intuitive awareness of, or opening out to, some giving entity or power that is bigger than oneself. In theistic terms, that entity would be God or some manifestation of God.

The theologian Richard Swinburne in his book *The Existence of God* (1991) identified various categories of religious experience. The first is the perception of God in natural objects. Wonderment at the night sky or at the structure of a rose might be among such experiences. Another category would include personal experiences that, although religious in nature, are describable in terms of common sensations. An example may be the experience of a dream in which one is addressed by a religious figure. Swinburne's next category is the personal response to witnessing what we have called extra-natural miracles. Then there are those personal experiences that are not describable in terms of common sensations, or perhaps not even in terms of sensations at all. For example, the mystics sometimes refer to a *spiritual sense* or of a sense of *nothingness* from which a feeling of unity with God is derived. Do such experiences truly reflect a direct interaction with the divine, or are they perfectly understandable as natural psychological phenomena? Perhaps the Victorians were correct in ascribing religious experience to an undigested pickle. This is the final evidentiary area we will consider.

So these are the evidentiary areas that will constitute the basis for the evidence denoted E in our Bayesian formulas. Through consideration of these areas, we will assess the probability that Proposition G is true. In the next chapter, the rubber hits the road and the numbers fly. Put fresh AAAs into the TI, sharpen the 2HB, and blow clean the Snoopy eraser: It's math time!

Chapter Conclusions:

Six evidentiary areas have been identified, and their bearing on Proposition G tentatively stated. These are

1. the recognition of goodness

2. the existence of moral evil

3. the existence of natural evil
4. intra-natural miracles
5. extra-natural miracles
6. religious experiences

seven

The Numbers

Questions for This Chapter:

What systematic process shall we use for sequentially updating the truth probability of Proposition G in light of each of our evidentiary areas?

How does each evidentiary area affect the truth probability of Proposition G?

The inconvenience of employing any kind of systematic approach to analysis is that it demands the establishment of a system. Let's begin by establishing *our* system. We have six evidentiary areas to consider:

1. the recognition of goodness
2. the existence of moral evil
3. the existence of natural evil
4. intra-natural miracles

5. extra-natural miracles

6. religious experiences

In due course we will identify the specific nature of the evidence in each of these areas, but for now we will denote that evidence as E1, E2, E3, E4, E5, and E6, respectively. Let's simplify our probabilistic terminology a little. Associated with each piece of evidence is a *before* probability and an *after* probability that Proposition G is true. The before probability is the preevidence, or *prior,* probability. This probability is then updated using Bayes' theorem, as the evidence dictates, to acquire the after probability (the postevidence, or *posterior,* probability) that Proposition G is true. We'll denote the before and after probabilities as P_{before} and P_{after}, respectively.

We will be considering the evidence serially; that is, we'll consider E1 then E2, continuing on to E6. For the first piece of evidence, E1, the before probability is P_{before} = 50%, this being our representation of complete ignorance, as discussed in chapter 4. That is, we ignorantly ascribe a 50-50 shot to God's existence. Based on evidence E1, we then use Bayes' theorem to acquire the after probability, P_{after}. This after probability then provides the starting point in consideration of evidence E2. That is, P_{after} for E1 becomes P_{before} for E2. Likewise, after Bayesian updating in light of evidence E2, the resultant P_{after} becomes the P_{before} for E3. This sequential analysis results ultimately in a P_{after}

from the last piece of evidence denoted E6. This final P_{after} is the answer we have been working toward: the probability that Proposition G is true.

(I'm compelled to interject an obscure point for the benefit of any Bayesian afficionados—it may be safely ignored by civilians. Note that we have considered the evidence items E1 through E6 to be mutually independent in the sense that $P(E1 \text{ and } E2|G) = P(E1|G) \times P(E2|G)$. This independence reflects the absence of any knowledge base by which to assume correlation between the evidentiary areas. That is, it is the least presumptive option. Now, back to the analysis.)

Let's now consider Bayes' theorem, introduced in chapter 4, as it applies to any one piece of evidence, denoted E for generality. So E could be E1, E2, E3, E4, E5, or E6. Our master formula is

$$P_{after} = \frac{P_{before} \times P(E|G)}{[P_{before} \times P(E|G)] + [(100\% - P_{before}) \times P(E|G^*)]}.$$

Looking at this equation, you can see that once we have the starting point of P_{before}, the two quantities we need to calculate P_{after} are $P(E|G)$ and $P(E|G^*)$. Let's briefly recapitulate on the meanings of these two quantities. They are:

> $P(E|G)$ = the probability that the evidence E would occur if G were true.
>
> $P(E|G^*)$ = the probability that the evidence E would occur if G were false (i.e., if G* were true).

Reiterating, $P(E|G)$ is the a priori probability that the particular evidence under consideration would in fact be before us if God exists, while $P(E|G^*)$ is the a priori probability that it would be before us if God does not exist. Now we must consider how these two quantities will be determined for each piece of evidence. In ideal applications of Bayesian inference, hard statistical data would form the basis for developing these probabilities. For example, in the case of the lie detector test and Le Mole discussed in chapter 4, the quantities $P(E|G)$ and $P(E|G^*)$ were based on the generic probabilities that the test produces false negative or false positive results. These probabilities would have been estimated through extensive analysis of the historic accuracy of the lie detection system. Proposition G has no direct analogy to this sort of statistical data, for this would demand knowledge of the properties of a multitude of universes, some godful and some godless. This type of information was unavailable at the time of publication. Therefore, here again we must draw on our Bayesian roots and rely upon a degree of judgment in assessing the quantities $P(E|G)$ and $P(E|G^*)$. That is, we will treat these parameters as Bayesian probabilities intended to reflect the strengths of the opposing theistic and atheistic arguments associated with each evidentiary area.

Now let's think a little more about the precise meaning of these quantities $P(E|G)$ and $P(E|G^*)$ in the context of the types of evidence we intend to consider. While we have identified specific evidentiary areas, it is clearly the case that our physical and biological universe, with all the particulars of its cosmological structure, physical phenomena, and life as we know it, might itself be construed as evidence of sorts. In a sense, our physical universe constitutes master evidence existing in conjunction with, or as a backdrop to, any specific areas of evidence we will consider, such as the existence of moral evil or extra-natural miracles. You can imagine that it might be difficult to frame our predefined evidentiary areas outside the specific context of the physical universe with which we are familiar. Well, perhaps if the universe had evolved somewhat differently and I were conducting this analysis as an eight-legged, two-brained Zard, or even as a living, intelligent complex of nuclear matter in the heart of a neutron star, the issue of moral evil might be just as germane to me. Nevertheless, to generalize consideration of our evidentiary areas outside the context of our specific, familiar world would be an impossibly difficult task. Happily, it's also an unnecessary task, as a consequence of certain arguments made in chapter 3, of which I will now remind you.

You'll recall that I spent some time considering the matter of whether the general physical and biological nature of our universe is a basis by which to assess the truth probability of Proposition G. I concluded not, thus rejecting what I called the rose petal argument and related perspectives. I refer you back to chapter 3 for the specifics of my case, but you may recall that it generally involved the anthropic principle, multiverse models, and shopping malls. A Bayesian expression of the conclusions I reached is that we really have no basis by which to assume that either $P(E|G)$ or $P(E|G^*)$ is greater than the other in the case that E is the evidence comprising general physical and biological phenomena. That is, there is no justification for assuming that the a priori probability of our particular universe is greater in either the case that G is true or that G* is true. So this piece of master evidence—our physical universe—is a wash, favoring neither G nor G*: it's God-neutral. The fortunate concomitant of this conclusion is that when we estimate the values of $P(E|G)$ and $P(E|G^*)$ for our specific evidentiary areas, we may do so against the background of our familiar world without biasing our analysis. So let the eight-legged Zards, with all their attendant problems of affordable footwear, find time to do their own study.

Now how will we estimate $P(E|G)$ and $P(E|G^*)$ for each area? First of all, let's take some advantage of the particular mathematical form of Bayes' theorem. It turns out that we have an opportunity to lighten our burden slightly by reducing the amount of information needed to put numbers into the formula. That is, we can do a clever little mathematical reorganization of the theorem to cast it as follows:

$$P_{after} = \frac{P_{before} \times D}{P_{before} \times D + 100\% - P_{before}}$$

where this new quantity D is defined by the equation

$$D = \frac{P(E|G)}{P(E|G^*)}.$$

From the formula for D, you can see that it is the factor by which a given piece of evidence is *more* likely to be produced in a godful universe than in a godless one. For example, if we state that D = 10 for a given piece of evidence, then we are saying that that evidence is 10 times more likely to be produced if G were true (i.e., if God exists) than if G* were true (i.e., if God does not exist). Let's call this D factor the Divine Indicator since the greater its numerical value, the more an item of evidence is assessed to point toward the existence of God—and the smaller its value, the more the evidence points away. The breakeven point is at D = 1, indicating evidence that is God-neutral; that is, P(E|G) = P(E|G*). The advantage we have thus gained is that we need now only come up with a single number, that is D, for each piece of evidence rather than the two numbers P(E|G) and P(E|G*).

For convenience, I will preset some representative numerical levels on the Divine Indicator scale so that we may refer to them as we consider each evidentiary area. Thus we are constructing a sort of scale that does for God-related evidence what Richter did for earthquakes and Fujita did for tornadoes. Our evidence scale is shown in Table D.

Table D presents definitions of a five-point Divine Indicator scale. This rather coarse resolution of the scale is a reflection of the somewhat approximate nature of the D estimates we will produce. That is, it would be tough (or tougher, more accurately) and indeed unrealistic to justify the

Table D The Divine Indicator Scale
(applied to each single item of evidence)

D Level	*Definition*
10	This evidence is *much* more likely to be produced if God exists than if he does not. (10 times more likely if G is true.)
2	This evidence is *moderately* more likely to be produced if God exists than if he does not. (2 times more likely if G is true.)
1	This evidence is God-neutral. (Equally likely to be produced whether it is G or G* that is true.)
½ (0.5)	This evidence is *moderately* more likely to be produced if God does not exist than if he does. (2 times more likely if G* is true.)
1⁄10 (0.1)	This evidence is *much* more likely to be produced if God does not exist than if he does. (10 times more likely if G* is true.)

selection of a specific value of D for certain evidence on, say, a 100-point scale given the nature of our considerations. In this system, we essentially allow the evidence to be assessed in one of the following ways:

1. It is much more likely to be produced if G (or, alternatively, G*) is true.

2. It is moderately more likely to be produced if G (or, alternatively, G*) is true.

3. It is neutral and therefore equally likely to be produced whether or not G is true.

By the phrase *much more likely,* I have selected the specific value of 10 times more likely or, in technical-speak, an *order of magnitude* more likely. Order-of-magnitude scales are commonly used in quantitative analysis where uncertainties and vagueness inherent in the subject matter render more refined scales unwarranted. The Richter scale of seismic activity would be a good example of such. "But," you might ask, "Why didn't he pick two or three orders of magnitude (i.e., factors of 100 or 1,000) instead of just one to represent the category of 'much more likely'?" Well, whereas some of you may wish to choose such large numbers in your own version of this analysis, in none of the evidentiary areas to be considered is my view so dogmatic and certain as to justify such large numerical differentials. That is, in each evidentiary area, I find the related arguments on both sides—both for and against Proposition G—to be reasonably compelling and in no case favoring G over G* (or vice versa) by a factor of 100 or more. So practically speaking, 10 will be the limiting value of the Divine Indicator for which I will have use. This point should become clearer once we begin to consider the evidentiary areas.

By the phrase *moderately more likely,* as defined on the Divine Indicator scale (see Table D), I have selected the specific value of two times more likely. There was clearly some freedom in the selection of this number, but I chose the lowest whole number consistent with preference for one of the competing propositions, G and G*. This point on the D scale will be used in cases where the evidence is assessed to lean in one direction rather than the other, but not excessively so, due to the strength of the competing argument. I think this numerical representation is consistent with the notion of evidence being moderately more likely in one case than the other. However, if these particular numerical selections cause you heartburn or are in any other way displeasing to you, the appendix to this book describes how you might assess the impact of choosing different representative points on the Divine Indicator scale. Again, use of this system should become clearer once we start to tackle the evidentiary areas. So let's proceed.

In the next six sections, I will summarize the key and relevant points of the six evidentiary areas identified in chapter 6 and then assess the numerical implications.

E1: THE RECOGNITION OF GOODNESS

In a godless universe, would *goodness* have meaning? Is it the presence of God's values in us that creates the asymmetry between good and evil such that we know we should do good and avoid doing evil? Or is that disparity borne of natural factors: evolutionary, sociological, or psychological?

Some perspectives on the origin of the good/evil asymmetry were outlined in chapter 6.

First, let me observe that *I* can discern generally between good behavior and evil behavior. Now in the real world there are many complex issues, the solutions to which lie in the gray. This grayness is generally borne of the need to weigh one good against another and one evil against another. Nevertheless, I believe that I can recognize the components of good and evil even in complex issues, though I may be unsure as to where the best compromise lies. So, for example, while many environmental issues are squarely in the gray, I understand that certain forms of atmospheric pollution will ultimately have evil consequence for future generations, while the near-term unemployment that might stem from certain remedial measures will also have evil consequence. Here, the grayness lies in deciding upon the appropriate balance of evils and not in the recognition of the elemental evils.

So why do I have this sense of good and evil? I know I consider things that hurt me to be evil in general. I am also familiar with the Golden Rule that one should "do unto others as you would have them do unto you." This rule makes sense since it is unlikely I am unique in regard to the types of things that hurt me. As the Dalai Lama repeats throughout his book *Ethics for the New Millennium* (1999), "All humans desire to be happy and to avoid suffering." So I try to follow the Golden Rule. Thus compassion is the pragmatic tool by which I act upon my recognition of the good. But what is my motivation for following the rule? If I do good unto person X, then there is an improved probability that person X will do good unto me. This will help me avoid pain and help *him* avoid future pain also. Thus we have a mutual understanding. But does acting on the recognition of goodness reduce to such pragmatic deals? Evolutionary models of moral behavior seem to imply so. Yet I often try to do good unto those from whom reciprocation is very unlikely. I give to charities, for example. I often give money to a homeless person without any anticipation of reciprocal favor. In these cases, I have no expectation of a return. I seem to have an attribute of compassion, and this in combination with a sense of empathy for others is what drives the will and ability to do good.

Now before I paint myself as Mother Teresa's kinder brother, I admit that there are instances in which, out of selfishness and convenience, I fail to do the good thing. Like most humans, I lie somewhere in that moral spectrum between Mother Teresa and a managed health care executive. Yet when I fail to do the good thing, the onset of guilt demonstrates that I know what the good thing would have been, even though I avoided doing it. If I were in the position of the person in need, I know what actions would aid me. I know what would reduce *my* pain, and compassion is the feeling that encourages me to reduce *his* pain. So this notion of compassion appears to short-circuit the pragmatic, mutual back-scratching view of goodness.

However, the argument has been made (see chapter 6) that pure pragmatism can also account for giving to those needy who lack the ability to reciprocate since in so doing we improve our own society. That is, in the long run *we* are the beneficiaries of compassionate actions since what goes around comes around. This would mean that it's still all about self-interest, although not always so directly as giving an apple to the teacher. I find this explanation a little difficult to accept. When I give to a homeless person, I lack any sense of anticipation that society will consequently improve. I see the value of my action purely as helping meet the needs, albeit temporarily, of that person. In a more extreme case, when I feed a stray dog, I have no sense of societal improvement—perhaps I even anticipate the opposite since a society ridden with stray animals is not necessarily a good one.

"Ah," you may say, "are your good actions effected to please the God you believe may exist?" If so, they reduce still to selfish pragmatism given the sorts of highly inconvenient things that certain scriptures promise for the ill-behaved. I'm thinking of the teeth gnashing and the hell program in general. I know certain people of faith who strongly believe that this sort of deterrent is the sole motivator of good behavior. This perspective would tend to give atheists free rein. For myself, I do not believe

Is the recognition of goodness, like politeness, learned?

that this prospect of comeuppance is the motivation for my wish to be good and behave morally. I certainly believe that atheists are no less prone to good behavior. I have as acquaintances many kind, compassionate people who will have absolutely nothing of religion, and so their goodness cannot be motivated by fear of hellfire. Of course, this is not to say that even their sense of goodness does not derive from God, whether they like it or not.

So does the fact that I have a concept of good and compassion that transcends pragmatic avoidance of personal pain imply that God is its source? Would this theory be dashed if there exist people who lack a sense of the good? An exchange from Reginald Rose's television play *Twelve Angry Men*, about the deliberations of a murder trial jury, comes to mind. One particularly brutish juror asks another juror why he is so polite, to which the second juror replies, "For the same reason you're *not*—it's the way I was brought up." Is the recognition of goodness, like politeness, learned? Are there people who are thoroughly bereft of the ability to recognize the distinction between good and evil because they were never taught how? But even if goodness is learned, it still requires an origin. Besides, I suspect it is not *recognition* of goodness that is learned but, rather, the inclination to *act* for the good—to view goodness as an important consideration in the making of decisions. This perspective would allow us to retain the notion that a sense of good and evil is inherent in us all. Otherwise, there would be parity between good and evil or, more likely, an utter absence of those concepts unless used merely to categorize one's own pleasures and sufferings.

Now, the naturalistic fallacy of G. E. Moore (see chapter 6), like the theistic perspective, dissociates goodness from evolutionary or sociological sources, yet in my mind, Moore's perspective does not fill the source gap. It offers no reason for the asymmetry in our perceptions of good and evil. So what does all this mean for the numbers?

Recall that the parameter we wish ultimately to quantify for each evidentiary area is the Divine Indicator, denoted as D. D is defined as follows:

$$D = \frac{P(E|G)}{P(E|G^*)}.$$

E, the evidence under current consideration, is that we occupy a world in which the distinction between good and evil is recognizable. $P(E|G)$ is the a priori probability that if God does exist, this evidence would in fact be before us. Well, if the person-God of the monotheistic faiths exists and he created us in his own image, then an inevitable consequence is that *we* would recognize the distinction between good and evil. After all, the good is a defining attribute of his character. So we must conclude that $P(E|G) = 100\%$; that is, we can state with certainty that inherent in our nature, as sentient occupants of God's universe, would be the ability to distinguish good from evil.

Now what of the parameter $P(E|G^*)$? This is the probability that if God does not exist, the current evidence would be before us. In a godless universe, would we inevitably lack the ability to distinguish good from evil? To assert so is to dismiss out of hand the alternative evolutionary, sociological, and philosophical models of morality. I would be unwilling to make that assertion. This unwillingness is rooted in acknowledgment of the possibility that I am being naive in assuming I can realistically diagnose the origin of my ability to discern good, moral behavior. While, for reasons already discussed, I find none of the atheistic explanations of compassion and morality highly convincing, it would nevertheless be overly zealous of me to assert that no future, natural explanations might be forthcoming or that current explanations might not be expressed more compellingly. Nonetheless, I am aware of no acceptable atheistic philosophical explanation for the genesis of the good, and the evolutionary models of moral behavior based on pragmatic, social principles seem inadequate. So since the evidence of goodness would be inevitable in a godful world, and its presence, at best, uncertain in a godless one, we can surely assert that this evidence is at least more likely to be produced if G were true than if G were false.

Now, how is this finding to be handled numerically in terms of quantifying the parameter $P(E|G^*)$? Say I express the conclusions of my analysis as follows: "The recognition of goodness is much more likely to occur in a godful universe than in a godless one." Looking at Table D, you'll see that this is precisely the expression of relative likelihood corresponding to a Divine Indicator value of $D = 10$. If we wished to select $D = 10$ for this evidentiary area, then since we have already established that $P(E|G) = 100\%$, this would be tantamount to stating that $P(E|G^*) = 10\%$; that is,

$$D = \frac{100\%}{10\%} = 10.$$

So assigning D a value of 10 is equivalent to the conclusion that there is only a 10 percent probability ($P(E|G^*) = 10\%$) that a godless universe would produce recognition of the good, thus allowing the existence of compassion. I would assert that $P(E|G^*) = 10\%$ is a reasonable numerical assignment reflecting the weakness of atheistic explanations. It is not an extremely small number (such as 1 in a million) since that would reflect incredulity at any sense of the good in a godless universe. Indeed, atheistic explanations cannot be dismissed and may ultimately evolve to produce more compelling models. So $D = 10$ is the appropriate numerical finding and the one with which we will proceed.

The analysis box for E1 shows the implication of these numbers. We begin with the preevidence probability of $P_{before} = 50\%$, which represents complete ignorance about the truth of Proposition G, and then update that probability in light of evidence E1. The result is that $P_{after} = 91\%$. Thus after consideration of the first evidentiary area, the truth probability of Proposition G has increased from 50 percent to 91 percent. A good start for God!

E1: The Recognition of Goodness

$D = 10$

$P_{before} = 50\%$

Apply Bayes' theorem:

$$P_{after} = \frac{50\% \times 10}{(50\% \times 10) + 100\% - 50\%}$$

So

$P_{after} = 91\%$

E2: THE EXISTENCE OF MORAL EVIL

Reconciliation of evil with the existence of an omniscient, omnipotent, and good God—the person-God of the monotheistic religions—is the chief bane of many people of faith. In chapter 6, some attempts at the reasoned reconciliation of evil with God were summarized. A principal theme is that in the long run, evil is in some sense a *good* thing, or at least a necessary . . . well, a necessary evil. Without the option of evil, our free will is trivialized: Without free will, there is no opportunity to choose the good. A related argument is that only under adversity do we grow spiritually and experience the most enriching aspects of life. So can these rationales be accepted against a background of events such as the September 11 atrocity, historic and contemporary genocides, random acts of severe violence, and other human horrors? It seems a callous explanation, but then perhaps our myopic human perspectives are not to be relied upon. Yet it seems that a role for evil is not a logical necessity since if the heaven we anticipate *does* exist, then it evidences a world solution that excludes evil. While we might lack the gall to demand a world that emulates the afterlife in this respect, would a little divine moderation of evil have thoroughly compromised the principle of free will? And besides, why is free will so much more important than the absence of evil? What good is inherent in a game that tests our inclination to make moral and good decisions? I have no answers, of course, so in accordance with our analytical philosophy, let's work the probabilities.

A principal theme is that in the long run, evil is in some sense a good *thing, or at least a necessary . . . well, a necessary evil.*

In a godless universe, moral evil is inevitable. Some people will pursue whatever means are necessary to achieve personal pleasure and gain, and, statistically speaking, some of those means will assuredly coincide with that which we consider evil. Therefore, $P(E|G^*) = 100\%$ is, I think, a reasonable assignment.

Now for the trickier question: Would moral evil exist in a universe created and minded by the omniscient, good, and omnipotent person-God? I'm inclined not to discount the divine rationales for moral evil that have been suggested by others. Perhaps the net good of life plus afterlife is positive for each of us, and the horrors of our earthly existence are ultimately more than offset. Perhaps the inability to make free choices is in itself an evil that God has protected us from. After all, we have knowledge of societies in which freedom to choose is excessively constrained, and the people of such societies are not, in general, happy folk. Nevertheless, on balance, I am inclined against the notion that the person-God would allow the extremes of evil that we witness and that the unfortunate experience. The free will argument has flaws, in my view, suggesting a lab-rat model of the human experience. Yet we experiment on rats only because there is a greater medical good. What is the greater good of God's experiment?

Again, there is always the ape and the monolith perspective, which would have it that I cannot hope to understand God's reasoning, if indeed *reasoning* be the appropriate word in the context of divine will. Well, perhaps evil does play an important role in allowing free will, or perhaps the divine rationale for evil is beyond human comprehension. Nevertheless, since $P(E|G^*) = 100\%$ and there is at least some uncertainty about the legitimacy of the theistic arguments for evil, one cannot but reach the conclusion that evil is more likely to exist in a godless universe than in a godful one. Yet the strength of the theistic arguments involving free will and the basis for opting meaningfully for the good leads me to disfavor the more extreme conclusion that evil is *much* more likely in a godless universe. That evil is at least *moderately* more likely in the case that G* is true is, I believe, an appropriate conclusion. This corresponds to a

E2: The Existence of Moral Evil

$D = 0.5$

$P_{before} = 91\%$

Apply Bayes' theorem:

$$P_{after} = \frac{91\% \times 0.5}{(91\% \times 0.5) + 100\% - 91\%}$$

So

$P_{after} = 83\%$

Divine Indicator value of 0.5 (see Table D), and this is the number with which we will proceed.

The analysis box for E2 shows the implication of these numbers. The probability that Proposition G is true in light of evidence E2 becomes $P_{after} = 83\%$. So after consideration of the second evidentiary area, the truth probability of Proposition G has fallen a little but remains solidly above the 50-50 level. Thus although God has sustained a minor blow, he maintains the upper hand as we move into the third area.

E3: THE EXISTENCE OF NATURAL EVIL

Earthquakes, tornadoes, cancers, droughts, airplane crashes, and other natural and technological hazards are the subject of this evidentiary area. Some of the principal issues here coincide with those raised under moral evils: Why would the omniscient, omnipotent, good God allow these natural hazards and their frequently horrific consequences? Divine rationales that have been suggested for natural hazards tend to be a little more limited than those put forward for moral evil since such hazards, by definition, cannot be construed as necessary concomitants of human free will. Nevertheless, the argument has been made (see chapter 6) that phenomena such as natural disasters and human illnesses can provide dramatic backgrounds against which our choice between good and evil becomes particularly poignant. In a sense, natural evils create for us a decision-making backdrop of Euripidean quality rather than that of some silly family sitcom. On such a stage our choice between good and evil has amplified significance.

While I would not dismiss this perspective, I find it less compelling than the corresponding arguments for the existence of moral evil. Whereas moral evil as a concomitant of free choice is an argument with some force, natural evil as a means of spicing up the human experience seems less convincing. Are human evils insufficient in degree and frequency to test us? Installing razor blades in the rat cage seems excessive. Once again, there is always the ape and the monolith perspective: Of course we don't get it . . . why should we? So the existence of natural hazards cannot be construed as a silver bullet, and indeed such may well be fully consistent with the existence of God. Yet I think the

E3: The Existence of Natural Evil

$D = 0.1$

$P_{before} = 83\%$

Apply Bayes' theorem:

$$P_{after} = \frac{83\% \times 0.1}{(83\% \times 0.1) + 100\% - 83\%}$$

So

$P_{after} = 33\%$

appropriate conclusion is that the existence of natural hazards is *much* more likely in a godless universe than in a godful one. Dispassionate, random nature is far more likely to dole out human misfortune in the circumstance that there is no compassionate God with the ability to intercede. So it is appropriate to conclude that this evidence disfavors God to a degree no less than the first evidentiary area, the recognition of good, favors him. Thus in reviewing Table D, we will pick a Divine Indicator value of $D = 0.1$.

The analysis box for E3 shows the implication of this number. The probability that Proposition G is true in light of evidence E3 becomes $P_{after} = 33\%$. So the third evidentiary area has done some damage to Proposition G, depressing its truth probability below the 50-50 mark to 33 percent. Now that G* has struck back hard, we suddenly have ourselves a game.

E4: INTRA-NATURAL MIRACLES

As defined in chapter 6, an intra-natural miracle is a difficult one to spot. It is an intervention of God that is effected through natural means, in the sense that there is no overt suspension of natural, physical laws. Perhaps the ultimate cause of the miracle, God's action, might be considered external to nature, but the balance of the process is not. The remission of an illness, the safe landing of a plane, the avoidance of a job layoff might each, under certain circumstances, be considered intra-natural miracles; but then under other circumstances, they may not. Each is an event that may be understood in terms of purely natural phenomena. Therefore, in what sense can an intra-natural miracle be identified as such? Some might take the extreme position that all events are intra-natural miracles since the universe exists and ticks by God's will. Well, okay, but then the concept of miracle becomes a trivial discriminator since all events are miracles. Let's adhere to the narrower definition that a miracle is a divine intervention that alters the future path of events.

Now, why would God adjust the course of events he has already set? Well, how about *this* rationale? Evidentiary area E2 addressed the notion of free will as the basis for the existence of moral evil. Sometimes that free will surely inclines us to reach certain points in our lives that God would not ideally have wished for us. Further, that free will causes us to desire certain outcomes to our earthly situations, even when we lack the power to directly bring them about. In these circumstances, we may simply harbor the desire for an outcome, or we may go a step further and pray. Where God responds to the desire or the prayer, that is a miracle. This is one basis by which we might presume to comprehend God's willingness to alter the preset course of events.

The remission of an illness, the safe landing of a plane, the avoidance of a job layoff might each, under certain circumstances, be considered intra-natural miracles.

Having loosely tied the notion of a miracle in this fragile bow, we may now ask the question of how miracles might be detected. Many of us can attest to the fact that not every prayer is answered (at least not as requested), and, presumably, desirable outcomes also happen in the absence of prayer. This would indicate the need for a statistical assessment of cases to determine whether prayer works. We might envision what physicists call a gedankenexperiment, meaning a "thought experiment." This is the concept of an imagined experiment that is not necessarily practical to perform. (An example might be dropping a stone on the surface of a distant star to assess its acceleration during fall: the kind of thing beyond the gall even of a government grant applicant.) In our gedankenexperiment, we might image two groups of people, each suffering from similar illnesses. Prayers are offered for the recovery of the individuals in one group, and the other unfortunate group remains exposed to the dispassionate rigors of nature. Would the beneficiaries of prayer reveal a statistically significant improved recovery rate?

You may be surprised to learn that for some people, *gedankenexperiment* is just another compound German word, and they have actually tried to perform this particular prayer experiment. Patrick Glynn reports on such experiments in his book *God: The Evidence* (1997). The results were inconclusive. I'm strangely comforted by this finding. Had one of the groups displayed significantly superior recovery rates, the conclusion might either have been that prayer is punished (since it represents greed? lack of faith in God's plan?) or that God is simply a service provider once we've done the equivalent of filling out the paperwork. Happily then, no hard data here.

Absent that data, I can resort only to my personal experiences. I have prayed, and not always for trivial, materialistic things. Furthermore, certain things have gone well in my life and other things less well. I admit to a certain psychological factor: When things go well, there is a tendency to attribute them to the results of prayer. When they go less well, the tendency is to assume that God has a better plan for me and those for whom I pray. On the other hand, in moments of cynicism, one is apt to dismiss prayer as an irrelevant act of despair borne of a sense of impotence. I have experienced all these feelings. The question I ask myself is, given my experience of prayer and outcome, what would be the a priori probability of this experience in (case 1) a godless universe and (case 2) a universe in which God exists? My intuition is to conclude that the a priori probability of my experience is greater in case 2. That is, my beliefs incline toward the notion that the results of prayer in my life have been, on balance, positive. By this I do not imply that I have perceived the results of prayer to be reliably and starkly beneficial. Yet, in hindsight, I have a sense of understanding the relationship between my prayers and my subsequent experiences. Now, I admit that there may be no impediment to this same perception in a godless universe also. I am not above self-deceit and indeed often rely upon it to get through the day. Purely random events often give the appearance of structure, as evidenced by the fact that technical stock market analysts continue to write newsletters.

Furthermore, even if God does answer prayers, is it necessarily the case that we would understand his response? After all, his reasoning is likely beyond our ken. Even if he exists, my perception of his interventions may still be self-delusional. Indeed, he may exist and yet not involve himself in prayer answering in the sense we assume. Yet, if there is never the potential to understand the relationship between prayer and outcome, would this not undermine our religious understanding of the person-God who encourages prayer? I believe that my sense of understanding the relationship between prayer and outcome is more likely, if only moderately so, in a godful universe. So I cannot dismiss the

E4: Intra-Natural Miracles

$D = 2$

$P_{before} = 33\%$

Apply Bayes' theorem:

$$P_{after} = \frac{33\% \times 2}{(33\% \times 2) + 100\% - 33\%}$$

So

$P_{after} = 50\%$

possibility that this sense of outcome is self-delusional and indeed cannot even conclude that the evidence is much more likely in a godful universe. Nevertheless, that my perceptions would be moderately more likely if G were true is, I believe, the appropriate conclusion. For this evidentiary area, I will set the value of the Divine Indicator at $D = 2$ (see Table D).

The analysis box for E4 shows the implication of this number. The probability that Proposition G is true in light of evidence E4 becomes $P_{after} = 50\%$. That is, the cumulative effect of all the evidence so far has been to balance out the scales exactly. So it's all up to the last two evidentiary areas. I, for one, am beginning to feel some sense of excitement.

E5: EXTRA-NATURAL MIRACLES

I have never witnessed an extra-natural miracle. The overwhelming odds are that you have never witnessed one either. If they truly exist, then they are rare. In our culture it is easy to be cynical about miracles. We are confronted almost daily by reports of UFOs, witchcraft, alien abductions, lake-dwelling behemoths, big-footed ape men, and fairies at the bottom of the garden. I think most of us feel relatively confident that there is no factual basis to such stories, and it is not difficult to view all reports of extra-natural miracles as part of this syndrome.

If such miracles do not occur, what would account for the reports? Deception and self-deception are the obvious explanations. Anyone who has enjoyed the performance of a professional magician knows that intellect is no impediment to being compellingly and completely deceived. Anyone who has read a tabloid while waiting in line at the supermarket checkout knows that there is no shortage of people anxious to fulfill the wish to be deceived. Then there are those who sincerely believe that they have witnessed a miracle. An ill-developed photographic image that reveals a figure where none had stood, crosses of light showing up in windows, and apparitions of the Virgin Mary are among typical reports. As someone who is fully seized of the diversity in the human condition, I would not even hazard to identify the psychological, sociological, or even biological reasons that someone would report a miracle where none had occurred. Whenever I presume to understand the rationale of a person who lacks that attribute of being me, I find a quick look at the newspaper comic strip page to be an immediate remedy to my arrogance. To see what I mean, try this. Pick up the comics section of a newspaper and select one of the comic strips at random. Read that strip through from beginning to end. Now realize that someone, somewhere is entertained by what you just read. This is a sobering reminder of human diversity and, in this case, diversity within just *our* society.

On the other hand, to dismiss all reports of miracles as part of the UFO syndrome would be an unjustifiably dogmatic position. So here's the approach I will take for the purposes of this analysis. I will decline to speculate as to whether or not authentic extra-natural miracles have occurred. I have no way of knowing the answer. Furthermore, I lack confidence that I could even design a credible means of assessing the legitimacy of extra-natural miracles given that a good magician can fool me completely at a range of six inches. Instead, the nature of the evidence I will consider is that extra-natural miracles have been, and continue to be, reported. Therefore, the question we need to consider is whether the very existence of such reports has impact upon the truth probability of Proposition G.

Miracles are a recurring theme in not only Judeo-Christian scripture but also in the mythologies of most

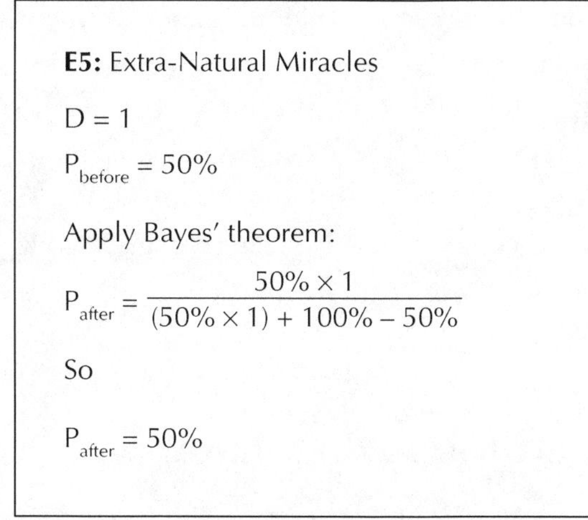

E5: Extra-Natural Miracles

$D = 1$

$P_{before} = 50\%$

Apply Bayes' theorem:

$$P_{after} = \frac{50\% \times 1}{(50\% \times 1) + 100\% - 50\%}$$

So

$P_{after} = 50\%$

theistic and polytheistic traditions, living and defunct. Miracles symbolize the enormous power of the subject divinities. Many followers of current religions believe in the existence of extra-natural miracles. This is an aspect of their faith. It is therefore inevitable, in my view, that reports of miracles will persist. That is, if miracle-believing religions exist, then so will reports of miracles. Now, miracle-believing religions have always existed, only some of which incorporate the God of our analysis. Zeus, Ammon, Bel-Marduk, and Odin, to name a few, were considered no more prone to respect the laws of nature. Therefore, it is inevitable that miracles will be reported regardless of whether G is true or false. That is, $P(E|G) = 100\%$ and $P(E|G^*) = 100\%$, and so the Divine Indicator, the ratio of these two probabilities, is simply $D = 1$. That is, the evidence is God-neutral.

The analysis box for E5 shows the implication of these numbers. The effect of a Divine Indicator of D = 1 (God-neutral) is to leave the truth probability of Proposition G unaltered. Thus it remains at $P_{after} = 50\%$ with only one more evidentiary area to go. At this point, I wish to apologize for any tension this mathematical exercise may be causing, but it does all seem to be down to the wire, doesn't it? Let's proceed.

E6: RELIGIOUS EXPERIENCES

I am uncertain as to whether I have had a religious experience. I have certainly looked at the night sky and felt a strong sense of awe at the sheer scale of the heavens. I believe my scientific knowledge of the universe's physical extent amplifies that sense of awe. That science can increase wonderment rather than reducing it is an idea that I find satisfying. Knowing that the distance to the edge of the visible universe is about 100,000,000,000,000,000,000,000 miles (that was 23 zeroes) should induce a sense of wonderment in any sentient being, but would it necessarily be a religious experience? I ask myself whether that experience of awe at the night sky is a feeling of oneness with the divine or of being opened out to a greater power in the sense expressed by philosophers such as William James and Richard Swinburne. I think for me that the answer must be "no." I suspect the reason for this conclusion is founded in the opinions I expressed in chapter 3 about the dissociation of matters cosmological from matters theological. Although scale is a bewildering aspect of our universe, it is not an attribute that I associate with the divine. Love of a tiny child—divine, perhaps; but awe at the bigness of emptiness—*not* divine, in my assessment.

Continuing on, it does nothing to suppress a feeling of spiritual inadequacy to note that neither have I experienced the most directly interpretable type of religious experience in Swinburne's taxonomy (see chapter 6)—that type involving the regular senses. These might be dreams or near-death experiences involving religious figures, tunnels of light, and visions of heaven. While I have dreamed about relatives who have passed away (and this type of dream is viewed by some as a religious experience), the context of such dreams was never a religious one.

I suppose I could choose the path taken for the evidentiary area of extra-natural miracles and define the evidence as simply the *reports* of religious experiences, which ultimately would tip the probability balance in neither direction. However, I am disinclined to opt for that path since, in this case, I find myself able to resort to my own experiences. Notwithstanding my sometimes drab, mechanistic outlook on the world, I believe that there have been occasions on which that outlook has been forcibly transcended. It is music that in my case, and on rare occasions, initiates a feeling that could be characterized in terms typically applied in the context of religious experiences—being opened out to a greater power. Such experiences come to me unexpectedly, and there is no reliable formula or set of conditions that dependably produce them. In a typical setting, I might be sitting in a church pew, transfixed on a stained glass window—not contemplating the significance of the scriptural image in the glass but simply absorbing the admixture of color and light. The most powerful enabler of the experience is the music, a Bach cantata, say. The resultant feeling is one of serenity, a brief absence of worry, and a sense that all is what it is meant to be. The feeling is wonderful, in contrast to my description of it. But is this a religious experience? Is having this experience related to the existence of the person-God? Well, as a Bayesian analyst, I will answer proudly that I don't know. Let's assess the probabilities.

Consider first a universe in which God exists. I believe that in such a universe, religious experiences of some type are inevitable. After all, if the person-God communicates with us personally and looks over us continually, it seems inevitable that, for at least some people at some time, his presence would be manifested in some direct and immediate way. This is not to say that I, specifically, would have an experience that could be characterized as religious. Yet, surely, during a lifetime in God's presence there would be at least one point in time at which he would be more directly manifested to me.

Now in a godless universe, could I have the sort of experiences I have described? Could they be fully explained as some natural psychological phenomenon, perhaps related even to a *need* for belief in something greater than myself?

After all, the enjoyment of music is something that need not be rooted in God. On the other hand, I *know* the experience of simply enjoying music, even to the point of spine-tingling pleasure, and the experiences I described are something different. Music is only one of the initiators of such experiences for me. Nevertheless, I cannot rule out the possibility that the experiences had no relationship to the divine, and thus I would be unjustifiably bold to conclude that $P(E|G^*)$ is equal to zero. Could that sense of something bigger out there be self-delusional, borne of acute desire for it to be true? Could it be a valid experience in the sense that there *is* something bigger out there, but that something bigger is something other than the person-God of our religions? Am I detecting the Spinozan God—the very essence of the natural world? I doubt this last possibility since I'm familiar with the thrill of experiencing Spinoza's deity. It's the excitement of scientific inquiry and the exhilaration of coming to comprehend an aspect of nature's subtle workings. That kind of experience is to be had with book or pencil in hand or behind the lens of a microscope. No, this is something quite different.

Whether or not such apparently religious experiences can be reconciled with a godless world, I would surely not be going out on a limb to conclude that they are *more* likely to occur in a godful one. However, to suggest that a godful universe is *much* more likely to produce this evidence would attribute insufficient credibility to the skeptical view that my experiences are not divine in origin but

E6: Religious Experiences

$D = 2$

$P_{before} = 50\%$

Apply Bayes' theorem:

$$P_{after} = \frac{50\% \times 2}{(50\% \times 2) + 100\% - 50\%}$$

So

$P_{after} = 67\%$

perfectly comprehensible in mundane psychological terms. Therefore, I will take the less dogmatic position that the evidence is only moderately more likely in the case that Proposition G is true. Hence, the Divine Indicator is assigned a value of $D = 2$ (see Table D).

The analysis box for E6 shows the implication of this number. The truth probability of Proposition G after consideration of the final evidentiary area now becomes $P_{after} = 67\%$.

And there you have it! Through systematic analysis of the evidence, I assess the final truth probability of Proposition G to be 67 percent. This means that the balance of probability—that is 33 percent (100 percent less 67 percent)—attaches to Proposition G*, which is that God does not exist. So comparing 67 percent to 33 percent, we have in effect assessed the odds at 2 to 1 in favor of God.

Your assessment of the evidence may differ. So now that you have the hang of the process, you may wish to adjust the numbers as you see fit and see what results you derive. You may even have new evidentiary areas to add. For the computer-inclined reader, the appendix to this book provides instructions on how to set up an electronic spreadsheet (Microsoft Excel, for example) to facilitate the calculations.

Now, as if knowledge of this magical number were not its own reward, we'll next consider how it might be applied.

Chapter Conclusions:

Some of our evidentiary areas point toward God's existence, and some point away.

Cumulatively, the evidence is in favor of Proposition G, although not overwhelmingly.

The truth probability of Proposition G is calculated to be 67 percent.

eight

Risk Apocrypha

Questions for This Chapter:

How might our newly calculated probability be applied?

How are probabilities generally used in decision making under conditions of uncertainty?

What is Pascal's wager on God?

How can we account for the prospect of an afterlife in probability-based decision making?

The probability that God exists is 67 percent. Well, let's not forget that this number has a subjective element since it reflects *my* assessment of the evidence. It isn't as if we have calculated the value of pi for the first time. Nevertheless, it's my best attempt, it was based on systematic analysis, and you know how I got there. The question now is, "How can I put this knowledge to use?" Well, what do we do with *any* probability number? Since

67 percent exceeds 50 percent, I can say that the preponderance of the evidence supports Proposition G. In a civil law court, this would generally be adequate to find in favor of Proposition G. So in an *Inherit the Wind* type of court setting, it's the proponent of Proposition G* that would go to jail.

Probabilities are used to make decisions under uncertainty. The 17th-century French mathematician Blaise Pascal had an interesting idea about the role of probabilities in religious decision making. This idea is sometimes referred to as Pascal's wager. We'll get to that shortly, but first I want to take a little time introducing some of the basics of probabilistic decision theory.

We begin by considering the simplest type of decision, one that involves no uncertainties. Say I am standing in front of two competing fruit stalls, $5 in pocket, with the intent of buying an apple. Fruitiverse Inc., the operator of the first stall, offers apples at $1 apiece. Java Applet, the adjacent coffee and fruit stall, offers identical apples at $1.25 apiece. Assuming all other things are equal, with which stall do I do business? I obviously choose to buy the apple from Fruitiverse since the benefit is greater. Buying from Fruitiverse, I will end up with an apple and $4 in my pocket. Buying from Java Applet, I will end up with an apple and only $3.75 in my pocket.

Of course, life is rarely a fruit stall and the vast majority of decisions involve uncertainties under which we cannot reliably predict the consequences of picking each option.

Enter decision theory. To introduce the basics of decision theory, let's temporarily return to the frequentist world of coin flipping. Say you and I have a fair coin and I make you this offer: I'll toss the coin, and if it lands heads up, I will give you $2; whereas, if it lands tails up, you will give me $1. Do you take the bet or decline it? Unless you're particularly averse to gambling, I think you'd take the bet. But why would you take it? Well, your expected gain from choosing the option in which you decline the bet is $0. That is a certain outcome. But what is your expected gain from choosing the option of taking this bet? If you win, you gain $2, but if you lose, then you lose $1, which we can call a gain of –$1 (that is, negative $1). Now in the apple purchasing example, I could compare gains directly and go with the option that produces the greater gain. Here, how do you compare the $0 gain of one option to the $2 or the –$1 gain of the other option? You need, in some way, to express the expected gain associated with the gambling option by some single number so that you can directly compare that number to the $0 gain of declining to gamble. This can be done as follows.

Imagine that the coin is tossed a large number of times and that you take the specified bet on each toss. Each time it lands heads up, you gain $2, while each time it lands tails up, you lose $1. If it's a fair coin, it will land heads up about half the time, and the greater the number of tosses, the closer will be the fraction of heads up to an exact 50 percent. Let's say it is tossed 1,000 times and it lands heads up 500 times (give or take). This means that your net gain will be:

$$500 \times \$2 + 500 \times -\$1 = \$500.$$

That is, after 1,000 tosses, you exit $500 up on the deal. So on average, the gain you made per toss was $500 divided by 1,000, which is 50¢. Returning to the question of whether you should take the bet, you now know that your average expected gain for a single toss is 50¢. This is sometimes referred to as the *expectation value* of the gain. Note that it is not the actual gain that would be made on any one flip since that is $2 or –$1. Indeed, the expectation value of 50¢ *cannot* be the gain on any single flip, but rather the expectation value is a measure of a hypothesized long-run average. This said, we will call the expectation value of the gain *the expected gain* for short. Reexpressing this concept in terms of flip outcome probabilities, we can say that the expected gain is

$$50\% \times \$2 + 50\% \times -\$1 = 50¢.$$

So now it is clear why you would take the bet. To decline the bet results in a $0 gain, whereas to accept the bet results in an expected gain for you of 50¢. In fact, say I wanted to charge you to make this bet. Now, the expected gain if you take the bet is 50¢ less whatever charge I impose on you to play. Provided that charge is less than 50¢, then your expected gain still exceeds the $0 gain guaranteed by declining the bet. This example captures the essence of a basic aspect of decision theory. To quantify the gain of a decision for which the outcome is uncertain, you consider each possible outcome; for each outcome, you multiply the gain by the probability of that outcome, then add these values across all possible outcomes. The result is the expected gain of that decision option. Then you do the same thing for each option and see which option has the greatest expected gain. All else being equal, this is the option you choose.

Now, although the probabilities in this example were based on coin-tossing concepts, decision theory is equally accommodating of purely Bayesian probabilities that lack direct frequentist interpretation. Let's consider another example before getting to Monsieur Pascal's wager on God.

In the previous example, there was uncertainty in outcome for only one of the two decision options. If you took the option of declining the bet, then the outcome was certain: nothing gained and nothing lost. Again, real life generally thrusts upon us situations in which uncertainty is rife and is involved in every available option. Let's look at a simple example of such a situation.

Kate attends the prestigious Chuck University. She has one more course to select this year and has narrowed down her decision to two options:

Option A: Introduction to Lie-Algebra-Based Differential Forms

Option B: Business Administration Part II

If she passes the algebra course, she will earn five course credits. If she passes the business course, she will earn three course credits. Failure of the course in either case earns no credits. She has more to gain by going with Option A and taking the algebra course, but here is the uncertainty: Will she pass the course? This is her thinking: The algebra course is known to be horrendously difficult. Even a friend of hers, who is able to solve second-order stochastic differential equations in her head, had been reported to weep upon seeing the final exam paper for this course last year. On the other hand, the business course had last been failed in 1988 when the exam paper had been accidentally placed upside down on a student's desk and he had read it as Cyrillic gibberish. Having passed Bayesian Inference 101 with flying colors last year, Kate determines the probabilities of passing the courses as follows:

> The probability of passing the algebra course she assesses as 65 percent. After all, her friend passed it and would be willing to help Kate. What's more, Kate has a good track record with math courses, and so she has a good chance of passing.
>
> The probability of passing the business course she assesses as 95 percent. She is reluctant to assign a larger probability because there is the risk that at some point in the course her eyes will glaze over and the crushing dullness of the subject matter will overwhelm her, precluding her ability to do the necessary work or indeed to maintain the will to live.

Therefore, the decision options are as shown in the analysis box.

Option A—Take the Algebra Course

Outcome 1:	Pass Algebra
Consequence of Outcome 1:	5 credits
Probability of Outcome 1:	65%
Expected Gain of Outcome 1:	65% × 5 credits = 3.25 credits
Outcome 2:	Fail Algebra
Consequence of Outcome 2:	0 credits
Probability of Outcome 2:	100% – 65% = 35%
Expected Gain of Outcome 2:	35% × 0 credits = 0 credits
Overall Expected Gain of Option A:	3.25 + 0 = 3.25 credits

Option B—Take the Business Course

Outcome 1:	Pass Business
Consequence of Outcome 1:	3 credits
Probability of Outcome 1:	95%
Expected Gain of Outcome 1:	95% × 3 credits = 2.85 credits
Outcome 2:	Fail Business
Consequence of Outcome 2:	0 credits
Probability of Outcome 2:	100% – 95% = 5%
Expected Gain of Outcome 2:	5% × 0 credits = 0 credits
Overall Expected Gain of Option B:	2.85 + 0 = 2.85 credits

So Option A has an expected gain of 3.25 credits, while Option B has an expected gain of 2.85 credits. It looks like the algebra course is the marginally better option from the perspective of risk. Note that some people are so averse to any uncertainty in life that for them this type of analysis would be moot and they would always opt for the alternative that has the most certain outcome. In this case, a mere 5 percent probability of failing business compared to a 35 percent probability of failing algebra would favor Option B. As a final comment on this example, note that our analysis is based on the assumption that Kate's sole interest is in course credits. Perhaps a more sophisticated analysis would look beyond credits as the only measure of gain and seek also to quantify other discriminants, such as sense of educational achievement or value of the course for the job market. However, our current interest in Kate's academic life has now expired, and we turn to God.

With these basic decision-theoretic concepts in hand, we will now set up the mathematical framework for explaining Pascal's wager. We begin by assuming that if the person-God exists, then there will be some afterlife in store for us (since all the monotheistic faiths maintain this position) and the nature of that afterlife is dependent, somehow, on the way we live our lives here on earth. Let us then proceed to assume that we have, generally speaking, two options for the way in which we live our lives:

Option A: To lead a God-pleasing life.

Option B: To lead a God-displeasing life.

These are very glib descriptors of the options and could have a broad range of meanings depending on your particular beliefs about God's pleasure. Later we'll discuss what it might mean to please God, but for now, suffice to say that if God exists, then the nature of your afterlife will be dictated by which of the two options you choose.

We will use the symbol P to represent the probability of the person-God's existence—that is, the probability that Proposition G is true. We have, of course, already calculated the actual value of P (as far as the author is concerned, at least). Now let's define some measures of gain:

$G_{please, god}$ is the reward (gain) of living a God-pleasing life if God exists.

$G_{displease, god}$ is the reward (gain) of living a God-displeasing life if God exists.

$G_{please, nogod}$ is the reward (gain) of living a God-pleasing life if God does not exist.

$G_{displease, nogod}$ is the reward (gain) of living a God-displeasing life if God does not exist.

Note that since these last two definitions relate to the case that God does *not* exist, then reference to a God-pleasing life might best be understood in terms of what the person-God

Option A—Lead a God-Pleasing Life

Outcome 1:	God Exists
Consequence of Outcome 1:	$G_{please, god}$
Probability of Outcome 1:	P
Expected Gain of Outcome 1:	$P \times G_{please, god}$
Outcome 2:	God Does Not Exist
Consequence of Outcome 2:	$G_{please, nogod}$
Probability of Outcome 2:	$1 - P$
Expected Gain of Outcome 2:	$(1 - P) \times G_{please, nogod}$
Total Expected Gain of Option A:	$P \times G_{please, god} + (1 - P) \times G_{please, nogod}$

Option B—Lead a God-Displeasing Life

Outcome 1:	God Exists
Consequence of Outcome 1:	$G_{displease, god}$
Probability of Outcome 1:	P
Expected Gain of Outcome 1:	$P \times G_{displease, god}$
Outcome 2:	God Does Not Exist
Consequence of Outcome 2:	$G_{displease, nogod}$
Probability of Outcome 2:	$1 - P$
Expected Gain of Outcome 2:	$(1 - P) \times G_{displease, nogod}$
Total Expected Gain of Option B:	$P \times G_{displease, god} + (1 - P) \times G_{displease, nogod}$

would have wanted *had* he existed. (After all, scriptures relate God's preferences either way.) Now we can derive the equations for the expected gain associated with each of the lifestyle options. These equations are shown in the analysis box.

So, for Option A (a God-pleasing life), we have

$$\text{Expected gain} = P \times G_{\text{please, god}} + (1 - P) \times G_{\text{please, nogod}}$$

while for Option B (a God-displeasing life), we have

$$\text{Expected gain} = P \times G_{\text{displease, god}} + (1 - P) \times G_{\text{displease, nogod}}.$$

Now we are in a position to define Pascal's wager in the language of mathematics. First, he implied in the description of his wager that the value of P is 50 percent. That is, he attached equal probability to the proposition that God exists as to the proposition that God does not exist. He noted that "God is, or he is not. But to which side shall we incline? Reason can decide nothing here. There is an infinite chaos which separated us." Like us, he expressed complete ignorance on the matter as a 50-50 toss-up. However, unlike Pascal, we used this 50 percent only as a prior probability, which we then updated using Bayes' theorem. No doubt Pascal is now kicking himself for not having thought of this in his *Pensées.* (Although he might be forgiven on the tenuous grounds of having died 40 years before the birth of Thomas Bayes.) Pascal also made some assumptions about the gains. He assumed that the gain of living a God-pleasing life, if God exists, is infinite. After all, we *are* talking about going to heaven here. So

$$G_{please,\ god} = \infty$$

where the symbol ∞ denotes infinity. Therefore, the expected gain of Option A is

$$\text{Expected gain} = 50\% \times \infty + 50\% \times G_{please,\ nogod}$$

and since half of infinity is also infinity, and any finite number added to infinity yields infinity (infinities really simplify math, don't they?), then the expected gain for Option A of living a God-pleasing life is

$$\text{Expected gain} = \infty.$$

As for Option B, who cares anymore? Assuming that the gain for living a God-displeasing life is not infinity, then the expected gain for Option B can only be finite (meaning not infinite). So the conclusion is that leading a God-pleasing life is the rational one because it results in infinite expected gain even in light of the uncertainty about whether or not God exists. Now Pascal noted that this argument is independent of the actual value of P, the probability of God. Even if P were assessed to be very small, the expected gain of Option A would still be infinite because even a very small number multiplied by infinity yields infinity—in fact, *any* finite positive number multiplied by infinity gives infinity. So Pascal noted that choosing Option A is always more rational than choosing Option B. That is, betting on God is always the wise thing to do. Is this an apocryphal notion of religious risk management?

Well, it seems like a reasonable argument. It boils down to the position that you should lead a God-pleasing life since if God does exist, then the reward is infinite; whereas, if he doesn't exist, well, what did you really lose? Is there any counterargument to this analysis, the results of which, in hindsight, are perhaps even a little obvious? More important, does Pascal really have no use for the actual value of P after we went to the trouble of calculating it to two significant digits?

It boils down to the position that you should lead a God-pleasing life since if God does exist, then the reward is infinite; whereas, if he doesn't exist, well, what did you really lose?

Some have found the time to voice criticism of the logic behind Pascal's wager. Following are some of those criticisms.

Zero Exceptions Except Zero

According to Pascal's argument, betting on God is always the best thing to do, regardless of the value of the probability P. However, there is one value of the probability P for which Pascal's argument breaks down, and that is when P is exactly equal to zero. Say you are pretty much a committed atheist but nevertheless have a minuscule shadow of a doubt about the nonexistence of God—that is, you assess P to be very, very, very tiny but not absolutely zero—then Pascal's argument still holds, and it is rational for you to live a life that is pleasing to God. However, if you are an absolute died-in-the-wool atheist for whom $P = 0$, then that's different. Although the P times infinity gain is equal to infinity for any nonzero value of P, when P is exactly zero, P times infinity is what mathematicians call an indeterminate quantity:

$$0 \times \infty = ??$$

This means that the result is undefined, but it is not infinity. This is a theoretically valid point, but as a practical matter, can someone truly assess P to be exactly zero—be absolutely certain that God does not exist? Well, there are those who profess this belief. However, given that there is a nonzero quantum mechanical probability that a giant kangaroo rat will spontaneously appear on your microwave oven, one has to infer the presumption of omniscience on the part of anyone who would set the probability of God to *absolute* zero.

An Infinity Is One Too Many

There are those in the decision theory community who balk at the notion of assigning infinite gains in any situation. They do not view this as a mathematically legitimate option. Is the heavenly reward truly infinite in value? For if it is only huge but not infinite, then Pascal's argument fails again. Assume that the gain $G_{\text{please, god}}$ is equated with an enormous number (say, a 1 followed by 100 zeroes) in some units of measure. This number is colossal, but it is not infinity. This would then imply that there is some breakeven probability (of God) below which Option A no longer offers the greater expected gain. This probability might be, for example,

P = 1/(1 followed by 100 zeroes).

This might make a difference for one of those virtual atheists who retains a tiny shadow of a doubt about the nonexistence of God. While even for virtual atheists Option A would be preferable if the heavenly gain is assumed to be infinite, Option B would become preferable below a certain threshold of assumed heavenly gain. Again, this is a mathematically valid criticism; nevertheless, I myself am disinclined to criticize the assignment of infinite gain to $G_{\text{please, god}}$ if the afterlife really does involve an eternity of happiness. But then maybe this is not the program for heaven. However, that uncertainty is beyond even my folly to analyze.

More Infinities, Please

Here's another thought that may compromise Pascal's argument. What if God is sufficiently merciful and compassionate to send us to heaven regardless of which option we took in life? In this case, both $G_{\text{please, god}}$ and $G_{\text{displease, god}}$ are infinite, both options have infinite expected gain, and the decision is moot. That possibility would upset someone you know, wouldn't it?

Another way to add an infinity to the analysis is to associate $G_{\text{displease, god}}$ with negative infinity. This would be hell, entered as a consequence of displeasing God. This, however, would not affect Pascal's argument. Option A would retain an infinite expected gain, but now Option B would have a negative infinite expected gain. Interestingly, the difference in expected gain between the two options would remain the same (that is, infinity) since

$$\infty - (-\infty) = 2 \times \infty = \infty.$$

So Option A remains preferable regardless of the value of P, and Pascal's argument still holds good.

Setting aside such mathematical criticisms, this whole decision analysis is based on the premise that we understand the nature of the decision options. It's all well and good to characterize an option as to lead a God-pleasing life, but what does this mean? Well, first of all let me confess that I have taken some liberty in defining the options as I do. Pascal had originally framed his decision alternatives with a focus on whether or not to believe in God. I have generalized them somewhat by using the God-pleasing language to capture the possibility that belief is not the criterion, or at least not the sole criterion, by which the heavenly turnstiles unlock. Each religion (or, more accurately, each branch of each religion) has its own beliefs about what it takes to get to heaven. These beliefs can be horrific (such as those held by the men who flew the planes into the Twin Trade Towers), discomforting, fair, unfair, sensible, or extremely silly, depending on one's world outlook. For those who believe there is some discrimination in determining who enters heaven, there are generally two competing types of beliefs:

1. Entry to heaven is determined by our behavior in life. That is, people who do good things for one another go to heaven. Of course, no one is perfect, so God's mercy plays an important role in the selection process.

2. Entry to heaven is determined by our religious beliefs in life. Certain Christians and Muslims, for example, are fully seized of the belief that their afterlife will not be plagued by those heathens whose beliefs in life differed even marginally from their own. For instance, those Christians who hold such a belief sometimes offer as one justification the passage from Ephesians 2:8–9, which reads, "For by grace are ye saved through faith; and not of yourselves: it is the gift of God. Not of works, lest any man should boast."

If the latter is accepted as the criterion for pleasing God, then we are stuck with the problem of how to create a belief in ourselves. After all, there is presumably no tricking God about our beliefs. Is belief controllable? Politicians would answer that question (in private) with a resounding "yes," but of course we as individuals lack the soft funds to shape beliefs. There is also the interesting question of the role of partial belief. Recall that partial belief is the very stuff of probabilities in the Bayesian outlook. Of course most sacred texts were written before the advent of Bayesian probability theory, and so they tend not to broach partial beliefs. It would be interesting to discover that the degree of joy in the afterlife is in direct proportion to one's degree of belief in certain religious ideas. What a vindication of the Bayesian interpretation of probabilities that would be!

So Pascal brushed much under the mat in establishing his wager on God. What *does* please God? Here is another uncertainty that might be addressed through Bayesian inference. Could we attach a numerical probability to each alternative religious theory of what gets us to heaven? My opinion on the matter, for what it's worth, is this. If God is the reason for our ability to distinguish good from evil and it is he who instills in us a sense of moral absolutes, then it is logically impossible for me to perceive God's judgment as less than perfectly moral and fair. However, I would view use of religious belief as the admission criterion to heaven to be enormously unfair on those brought up in the other errant faiths. So this cannot be the way God behaves. *Q.E.D.*

It would be interesting to discover that the degree of joy in the afterlife is in direct proportion to one's degree of belief in certain religious ideas. What a vindication of the Bayesian interpretation of probabilities that would be!

Notwithstanding these uncertainties and ambiguities, I see some validity in the conclusions of Pascal's wager. No matter how small the probability of God's existence, behaving as if he does exist is the smart thing to do. But since this conclusion is independent of the actual value of the probability, provided that it is not zero, is knowledge of a specific numerical probability helpful? Does P = 67% imply, for example, that I should attend church only 67 percent of Sundays or that I should reduce my collection plate offering by 33 percent? No, that's crazy frequentist talk.

I think there is value in having in mind a sense of the likelihood that God exists. A value of 67 percent has very different implications in my mind from a probability of one in a billion. Whereas Pascal sees no distinction (wager-wise) between 67 percent and a one-in-a-billion chance, my sense of imperative to please God is more sensitive to this probability. That the probability is greater than the preevidence prior of 50 percent also has some significance for me. While the probability swung back and forth as we considered each evidentiary area (from as little as 33 percent to as much as 91 percent—although the values of these intermediate extrema depended on the order in which the evidentiary areas were considered), the final probability exceeded the initial prior probability of 50 percent, which you will recall was intended to represent complete ignorance on the matter. That is, the net impact of the evidence was in favor of God—but not excessively.

Furthermore, I see merit in knowing the result of a reasoned analysis that seeks to capture in a nutshell the yield of having thought through the evidence to the best of one's ability. It is as if the scattered pebbles of one's experiences and perceptions of God have been gathered and placed neatly in a box. Maybe the box of pebbles has no overt, direct use the way a can opener does, but I think there is value in having cleaned up the mess of pebbles. As new pebbles are found, they can be added to the box. By this I mean that new evidence or reinterpretations of existing evidence can be used to maintain the truth probability of Proposition G as an evergreen representation of my beliefs. So much for pebbles.

I'll finish this chapter with a question. Is the result of 67 percent consistent with my intuitive feel for the probability that God exists? That is, is it about equal to the truth probability I might have produced off the top of my head had I bypassed this Bayesian analysis? Well, that question leads us to the matter of faith and perhaps to the real significance of the 67 percent.

Chapter Conclusions:

When one is confronted by a range of decision options, each of which has uncertain outcome, probabilities help order the options (best to worst) with respect to their anticipated, average benefits.

Pascal's wager applies basic decision theory to the question of whether or not one should believe in God (or live a God-pleasing life), balancing the implications of each option for one's admittance into the afterlife.

Pascal's arguments lead to the conclusion that it is rational to believe in God regardless of whether or not he exists.

Probabilities have intuitive significance beyond their application in formal decision theory.

nine

Probable Thoughts

Question for This Chapter:

What are the perspectives of Chad and The Axe, the mall intellectuals?

THE PROBABLE THOUGHTS VIGNETTE

Scene: A shopping mall fast-food restaurant adjacent to the Body-Part Piercing Parlor. The two mall boys are sitting at a table with food before them.

Characters: The mall boys—Anaxagoras (The Axe) and Chad.

[CHAD *is hunched over the table holding his nose.*]

CHAD: This really hurts.

THE AXE: The swelling'll go down soon.

CHAD: But this *really* hurts. I think I'm hemorrhaging.

THE AXE: If you were hemorrhaging, that fluid you're oozing would be red, not translucent green.

CHAD: Do you think those people are really qualified to do this?

THE AXE: At the Body-Part Piercing Parlor?

CHAD: Yes.

THE AXE: Oh, sure. I think I saw a Harvard medical diploma on the wall.

[*A silence, except for anguished sniffling.*]

THE AXE: Chad, are you a man of faith?

CHAD: Faith? What do you mean?

THE AXE: You know. Religion. Do you believe in God?

CHAD: Yes, I do.

THE AXE: Oh, really.

[*A short silence.*]

THE AXE: I've just completed a little project. I've calculated the probability that God exists.

CHAD: How did you do that?

THE AXE: Well, I lined up the evidence and then used something called Bayesian inference to systematically calculate the probability that God exists. I doubt that you're familiar with Bayesian inference.

CHAD: I've heard of Bayesian inference.

THE AXE: Oh? It's not often you hear that from a man oozing green slime from his nostrils.

CHAD: What was the result?

THE AXE: 67 percent.

CHAD: Hmm. Interesting. I can't eat this burger. Do you want it?

THE AXE: No, I don't like the green toppings. [*Pause.*] So I'd say that that probability is pretty high. Even higher than break-even. A good reason to have faith in God's existence, don't you think?

CHAD: What do you mean?

THE AXE: Well, when I began this calculation, I had no idea what the result would be. It could have been a probability of one in a thousand, a 50-50 shot, or 99 percent. It happened to wind up in the middle ground, which seems to be a reasonably good chance for God. I'd say that means faith in God isn't ridiculous.

CHAD: I think I see your reasoning. So you're saying that the justification of faith is in direct proportion to the reasoned probability of God's existence.

THE AXE: [*A little stunned by the acuteness of* CHAD*'s understanding.*] Yes. I guess that's what I'm saying.

CHAD: You see, I don't view faith that way. I think that for a believer, faith is the thing that bridges the gap from reasoned probability to complete belief.

THE AXE: What do you mean?

CHAD: I know that nothing is real in your eyes until it can be scribbled as a math formula. Pass me that napkin.

[CHAD *writes down:* $F = 100\% - P$]

THE AXE: What's that?

CHAD: P is the result of the calculation you did—it's the probability that God exists, or, in Bayesian-speak, your degree of belief that God exists. Then F is the measure of faith required to achieve full belief in God. That is, add F to P and you have 100 percent belief. So you say that your value of P is 67 percent. That means to fully believe in God, your faith must supplement P by 100 minus 67 percentage points, which equals 33 percent.

THE AXE: So you're saying that faith is complementary to reasoned probability in achieving full belief?

CHAD: That's what *I* think.

THE AXE: Then, in fact, the greater one's reasoned probability, the less the role of faith since faith has the function of eliminating the residual uncertainty about God's existence.

CHAD: Exactly.

THE AXE: But that makes faith something that's contrary to reason, instead of being produced *by* reason.

CHAD: I suppose that's right.

THE AXE: So is faith good or bad?

CHAD: I don't think that faith is inherently good or inherently bad. Faith in good things is good, faith in bad things is bad. Those guys who flew jet planes into the World Trade Center had faith. My mom has faith that I'll pass American history this semester. I wouldn't lump those faiths together as either good or bad. I'll tell you this, though: Faith is always dangerous.

THE AXE: Why?

CHAD: Well, by its very nature, it's impenetrable by reason. In fact, as the complement to reason, it may grow with mounting counterevidence. So faith is very powerful, and you have to hope that where it exists, its object is something good.

THE AXE: This seems wrong to me. Say I tell you that the universe was laid by a giant magic toad. That's obviously a preposterous creation theory to which the assigned truth probability would be minuscule. But according to your formula, this means that the faith factor would be very high.

CHAD: Only if you fully believe in the toad genesis story. If you think the story untrue, then you have no desire to bridge the gap from reasoned probability to full belief.

THE AXE: I see. But still, none of this sounds right to me. What's the origin of faith? Reason has some universality. People in Finland and Fiji agree on the rationality of arithmetic. But there's no common denominator to faith. You have faith in X—end of story. Surely great minds over the ages have put forth reasoned arguments for the existence of God. Many of these smart people were people of faith. Even now, debates about intelligent design of the world are prevalent, and rational analysis is being used to create a case for God. Why would they wish to build a reasoned case for God if the effect is to reduce the role of faith?

CHAD: Well, I'll tell you what *I* think, for what it's worth. There *are* many bright, accomplished people who are people of faith. I think that it's in their nature to be analytical and to seek to uncover rational justification for their beliefs. So they apply their intellectual resources to the task and often produce articulate, structured cases for God, the way they might produce a case for a scientific theory, a business plan, or an engineering method. But while their cases have all the trappings of rational, defensible analysis, it's largely superficial. They entered their analysis with faith in the existence of God. The conclusions of their analysis were thus hardwired from the outset, and they merely sought an articulate, credible route to the predetermined end point. They may import the vernacular of reason that is so successfully applied in other fields, but the difference is, in those other fields they do not start with an immutable prebelief. Imagine a dentist who, before his patient opens her mouth, decides that it will be the upper-left canine that's the problem. So the process of justifying faith is in my opinion a very artificial one.

THE AXE: But I've heard the criticism that many scientists work that way—they have a theory and bend the facts to it.

CHAD: Yes . . . and that's bad science, and scientists recognize it as such. However, people of faith seldom rebuke each other for getting to their preestablished end point though *reasoned* argument. I think that since their beliefs were not arrived at through reason, it's disingenuous of them to use the instruments and language of reason to retrofit rationale to those beliefs.

THE AXE: Wouldn't you accuse atheists of the same thing? They'll enter into an assessment of the evidence for God having predecided their conclusions.

CHAD: I agree. They're equally disingenuous. It's the subject area itself that mocks reason. Both sides of the argument are guilty.

THE AXE: Am I guilty of this prebelief in *my* analysis?

CHAD: I don't know where you started from. I must say that the nature of your objective was rather novel—you know, calculating the probability of God. I doubt that you had a preestablished number in mind that you worked toward, but you probably did have at least some degree of belief and skepticism when you engaged in your analysis. However, I suppose your objective was simply to represent and analyze that combination of partial belief and partial skepticism in a systematic framework to assess the probabilistic implications of your views. Maybe that's the beauty of working with Bayesian probabilities—their subjectivity is overt, yet they facilitate and make more transparent your reasoning process.

THE AXE: Well, I thank you for *that*. But I still think you're being unfair, Chad. A belief in God doesn't obviate someone's ability to rationally consider the evidence for God's existence. Can't the intellectual part of the brain do its work in isolation from the faithful part?

CHAD: I really don't think so. Or at least I've never seen evidence that it's possible. I think that faith, by its nature, invariably trumps reason. And perhaps where intellectual compromise is self-recognized in people of faith, they see it as a small sacrifice. I don't know. Perhaps a purely reasoned view of the world would be just too dreadful, and taking a few liberties with reason is no big deal in the scheme of things.

CHAD: By the way, now that you have your result, tell me, do you have enough faith to bridge the gap from your 67 percent probability to full 100 percent belief?

THE AXE: If I don't know the answer to that, does that mean the answer is really "no"?

CHAD: If you can't answer the question, maybe it just means that for now you are still uncertain about God. Whether your faith supplements your reasoned probability by only a few percentage points or whether it takes you to within a hair's breadth of 100 percent belief, only you know. Can you calculate faith?

THE AXE: Chad, do you think it's legitimate to have faith in the virtues of behaviors promulgated by the major religions without believing in God? Would, for example, the Christian message of compassion be any less valid without a God to back it up? After all, the Buddhists teach those same virtues, and they don't seem to need a deity. Perhaps someone who lives the Christian or Jewish or Islamic life benefits regardless of whether God is there. I often envy the great sense of contentment that many people of strong faith seem to enjoy.

CHAD: For me, God *does* make the difference. I guess I believe that my sense of a good life *derives* from him. So to speculate on doing good without his existence is meaningless for me. And maybe it's also that I'm a little shallow. Perhaps I like the notion of a pat on the head for doing the right thing. Perhaps it's a sense of the repercussions of evil that discourages me from doing it. But mostly, the thought of a godless universe is just too horrible for me to contemplate.

THE AXE: Chad, you have stunned me today.

[CHAD *smiles, holds his nose with one hand, and carefully bites his hamburger.*]

CHAD: When we're finished, I'd like to go back to the "You Are Here" arrow on the shopping mall guide. I can't explain why, but I like to look at it.

THE AXE: Mind if I tag along?

CHAD: Let's go.

Curtain falls.

Chapter Conclusion:

Think carefully before having your body pierced.

ten

Faith Math

Questions for This Chapter:

What is faith-type belief as it applies to specific propositions?

How do faith-type belief and partial belief (as represented by probabilities) differ from one another?

How do faith and reasoned probability combine to form an overall level of belief in the truth of a proposition?

Can an atheist experience faith-based belief?

By now you should be fully indoctrinated in the idea that beliefs are related to probabilities. This idea is the essence of the Bayesian school of probability theory. More specifically, a probability is a measure of partial belief, or level of confidence, in the truth of a proposition. Now surely, faith is also related to belief in certain propositions. Well then, if both faith and probability reflect elements of belief, it seems natural to ask the question of how faith and probability are related to one another. Is there some linkage between the two concepts that will allow us to leverage our understanding of probability to better comprehend the nature of faith? Or is the relationship between the two largely superficial and based on no more than a tenuous semantic connection such as that between warm pizza and warm feelings? Now that we've emerged from our Bayesian adventure with heads full of notions probabilistic, I think the time is propitious to take the plunge and see what light might be shed on the matter of faith, particularly as it bears upon specific propositions.

Let's note first of all that the word *faith* is used generally in two very distinctive contexts. First is the objective version of the word referring to a system of beliefs. Thus a person may be of the Islamic faith or Catholic faith. However, it is the other, subjective application of the word that is the focus of our interest. This faith is related to the concept of an individual's belief. But what is the form and origin of such belief?

If both faith and probability reflect elements of belief, it seems natural to ask the question of how faith and probability are related to one another.

One perspective on the relationship between probability and faith is that whereas the intent of the former is to represent *reasoned* belief, the latter necessarily represents *unreasoned* belief. That is, faith, in contrast to probability, is not derived from the analysis of evidence and, worse still, is completely disconnected from reasoned thought. This is generally the atheistic notion of faith. While this might be a tidy classification system for belief types, is it really fair or even valid? As you might imagine, there are many contrasting viewpoints on the nature of faith, and we shall get to some of them shortly.

To begin our deliberations, let's investigate Chad's theory of faith, as introduced in chapter 9. Chad has not given excessive thought to the matter of faith, and so our expectations of his theory's cogency should be modest. Nevertheless, it's a simple and elegant theory that might help establish some discussion points.

Chad observes that many people are certain of God's existence and many are certain of his nonexistence. Let's assume for now that when someone professes such certainty, they are being honest. But surely, reasoned analysis cannot be the sole source of this certainty unless one hypothesizes serious flaws in the reasoning apparatus available to the believer. Indeed, believers in God, for example, would not generally claim that their complete belief derives from reasoned analysis. Therefore, Chad views faith as the thing that removes residual uncertainty after reason has done its best. That is, faith supplements reasoned partial belief, topping off the belief tank and bringing the needle gauge to the "Full" mark. We can express this idea mathematically as follows. Remember that P(G) is the truth probability of

Proposition G. That is, P(G) is the probability that God exists based on our reasoned analysis. Let's now denote as F(G) the faith-based component of belief that God exists. Chad's theory has it that for a believer in God, the following equation is true:

$$P(G) + F(G) = 100\%. \qquad \text{(Equation 1)}$$

That is, you start with reasoned partial belief (probability) and then supplement it with the faith factor to achieve full 100 percent belief. This seems like a very tidy, succinct little formula, but let's explore its implications a little further before we submit it to the *Encyclopedia Mathematica* to take a hallowed place between $E = mc^2$ and the formula for compound interest.

Recall that we have been dealing with two propositions, denoted G and G*, one being the negation of the other:

Proposition G = God exists.

Proposition G* = God does not exist.

These propositions are, together, both of the following:

- exhaustive—that is, in combination, they cover all possibilities.
- mutually exclusive—that is, only one of them can be true, and the other is then false.

Another example of two exhaustive and mutually exclusive propositions applies to the toss of a coin. These propositions are "The tossed coin will land heads up" and "The tossed coin will land tails up." That is, these propositions are exhaustive because there is no other possibility but heads or tails (assuming that the coin will not land on its edge), and the propositions are mutually exclusive because only one of the sides will land facing upward. If we ask what the probability is that the coin will land *either* heads up or tails up, the response would be 100 percent. That is, there is no third option because the propositions are exhaustive. Furthermore, because the propositions are mutually exclusive such that the outcome must be heads alone or tails alone, we can state that the probability of heads plus the probability of tails is 100 percent. Similarly, with Proposition G, we can write down the formula:

$$P(G) + P(G^*) = 100\%. \qquad \text{(Equation 2)}$$

Or in words, the probability that God either exists or doesn't exist is 100 percent. This must seem to you an obvious statement. But now compare Equation 1 with Equation 2. These formulas are identical except that one contains an item F(G), while the other contains an item P(G*). If both these equations are valid, then this implies that these two items are identical to one another. That is

$$F(G) = P(G^*).$$

So what does this formula say? It says that the faith-belief attached to Proposition G is equal to the reasoned belief attached to Proposition G*. Put another way, the faith factor attached to the proposition that God exists is equal to the reasoned probability attached to the proposition that God does not exist. This seems like an unsettling implication of Chad's theory. That is, the greater the reasoned belief and evidence *against* God, the greater the faith *in* God. Now it has become clear that although at first glance Chad's theory of belief appears to be rather benign, in fact it solidly implies that faith is the antithesis of reason. Thus a little math has exposed that if Chad's theory is correct, then faith is blatantly irrational. This theory would imply that faith is not only a form of belief disconnected from evidence but also something much worse: that faith grows with counterevidence. This seems to be an unsatisfactory notion of faith. It's one thing to assume that faith and reasoned belief have their origins in different parts of the brain, but it seems to be going too far to assert that faith requires us to actually reverse-wire the machinery of reason.

Perhaps there is a way to retain the general notion that faith in some way supplements reasoned belief without leading ourselves to the sorts of perverse conclusions implied by Chad's theory. What are the offending assumptions of this theory that perhaps we can discard or relax a little? One critical assumption is that faith always serves to completely remove any residual uncertainty about the truth of Proposition G. This is the assumption that sweeps us uncontrollably to the undesirable conclusions of the theory. That is, by requiring faith to top up belief to the 100 percent level, the concomitant is that faith is mathematically equivalent to antireason. But is it necessary or even appropriate to assume that a believer in God must achieve a belief level of 100 percent? In their book *Life's Big Questions* (2002), William Grimbol and Jeffrey Astrachan suggest that

> Faith without doubts is blind faith. . . . Without doubts, faith becomes nothing more than a religion that calls for the removal—not the renewal—of your mind. A faith void of doubt is the product of indoctrination, not inspiration.

In the Grimbol-Astrachan view of faith, doubt is not merely to be tolerated, but it is a desirable element of belief in God. Doubt presumably indicates that the belief resides in a living, vital mind and not in one deadened or at least anaesthetized by indoctrination. This perspective allows us to be a little less dogmatic with our math. Perhaps another less restrictive mathematical formula involving the faith factor is warranted. What if we posited that the total degree of belief in Proposition G, denoted B(G), has two contributing components: reasoned partial belief, denoted P(G), and the faith factor, denoted F(G). The belief formula could then be written as

Doubt is not merely to be tolerated, but it is a desirable element of belief in God. Doubt presumably indicates that the belief resides in a living, vital mind and not in one deadened or at least anaesthetized by indoctrination.

$$B(G) = P(G) + F(G).$$

Now, while Chad's theory demands that P and F for a believer must add to produce $B = 100\%$, *we* will no longer make that assumption. Indeed, consistent with the perspective of Grimbol and Astrachan, B should fall short of 100 percent by some doubt factor. But how large a doubt factor is consistent with overall belief in God? Could we have zero faith, that is, $F(G) = 0$, and consider the doubt factor to be the amount by which our reasoned probability P(G) falls short of 100 percent? So if $P(G) = 67\%$ (as in the case of the author), there would be a doubt factor of 33 percent. My intuitive feel is that this would be an excessively high doubt factor for a believer. Therefore, the faith factor would be badly needed to pick up at least some of the slack. We might expect faith, denoted F(G), to be a quantity that adds to reasoned probability, denoted P(G), to bring the total belief factor of a theist to within one, two, or perhaps a few percentage points of 100 percent. So

$$B(G) = 98\%$$

might be a reasonable, typical representation of a theist's aggregate belief in the truth of Proposition G. In this case, the doubt factor is only 2 percent. So we now have a simple little mathematical model defining the way in which reasoned probability and faith add to form the overall level of belief in the truth of a proposition. What's more, by dropping the requirement that the belief of a theist in Proposition G must be a full 100 percent, we have obviated the nasty little implication of Chad's theory that faith is equivalent to antireason. Well, with this degree of elegance in a theory, any red-blooded academic would by now have published 15 papers, have presented their findings at eight international conferences (and like all international conferences, strangely, each attended by the same 38 Scandinavians), and be demanding doubled research monies and halved teaching duties. But not *me*. I'm going to take the cautious and responsible path of investigating this theory for a little while longer. What, for example, is the significance of this theory as it applies to atheists?

Is there such a thing as the faith of an atheist? Is it appropriate to assume that an atheist's beliefs do not consist exclusively of the reasoned element? Let's presume that even an atheist would, based on reasoned analysis, produce some nonzero probability (albeit very small) that God exists. This would imply that an atheist's assessed probability that God does not exist, denoted $P(G^*)$, would be large but would nevertheless fall short of 100 percent. Is it therefore faith that leads the atheist to declare 100 percent certainty about the nonexistence of God? Does it make sense to express atheistic belief by the analogous formula

$$B(G^*) = P(G^*) + F(G^*).$$

In this case, B is the total level of belief in the nonexistence of God, P is its reason-based component, and F its faith-based component. Or is it purely semantic mischief to try and assert some symmetry between the theist and atheist regarding the role of faith? Is P(G*) so great for the atheist that F(G*) would be a quite unnecessary, and perhaps even meaningless, addend? It seems we can go only so far with this math business before we need to explore in a little more depth the nature of the faith factor I have glibly labeled "F" in the formulas.

Is there such a thing as the faith of an atheist?

Returning to the original strawman position, is it fair to characterize faith as the reasonless sibling of probability? This perspective would have it that faith is associated with credulity, blind and unquestioning belief, or belief without evidence. The contemporary American Bible scholar Doy Moyer recoils from this perspective. For him, faith is something that is squarely founded in reasoned analysis, as he explains in his *Focus* magazine article "A Study of Faith." Moyer views the evidence for God, and more specifically for the validity of Christianity, as being essentially historical in nature. Although he asserts that faith should not be viewed as a leap in the dark, he nevertheless acknowledges that faith often requires a willingness to

> look a little beyond the evidence, all the while coming back to the solid ground that evidence provides.

So, for example, he notes that

> The historical events of the death and resurrection of Jesus, for which there is much evidence, serve as the foundation for accepting the idea that Jesus died for our sins.

Moyer appears to be saying that if we accept the historicity of the life and resurrection of Jesus, then it is merely looking "a little beyond the evidence" to conclude that Jesus died for our sins. Does this extension of belief to the rationale for Jesus' death really constitute no more than looking a little beyond the evidence? But then again, if faith *is* more than just looking a little beyond the evidence, does this really compromise its substance? My opinion is that it does not. In fact, the idea that faith is a cautious, limited excursion beyond the boundary of evidence is one that I think inappropriate. This type of language may apply to the concept of a scientific conjecture, but surely faith allows us to travel further afield than the easement of the evidence.

There are other notions and explanations of faith with which I am equally uncomfortable. Some can verge on the villainous, particularly where they are motivated by the perceived threat of science to faith. For example, one is sometimes confronted by the equalizer perspective. It goes something like this:

> When your teacher tells you that the distance to the moon is a quarter of a million miles or that there are two protons in a helium nucleus, you have no way of verifying it, do you? So you take it on faith. That's no different from having faith in a religious worldview.

This argument can then be extended to any scientific or even historic information. Thus the waters are effectively and deeply muddied. The implication is that an equivalence exists between scientific teachings and religious teachings since each requires faith in the teacher. This is, of course, a treacherous argument based on tricky semantics. We can certainly use the word *faith* as a synonym for trust in the information provided by a credible, credentialed source such as a teacher. However, this variety of faith is essentially a rational belief that the information conveyed by the teacher originated initially in the reasoned analysis of evidence, whereas the very information that is being conveyed in religious teaching is often (although not always) a manifestation of a faith-based belief. That is, the equalizer argument relies on faith-fuzzing the distinction between the messenger and the message.

So here are two perspectives on faith with which I cannot go along: that faith involves only a minor excursion beyond evidence and that faith is no more than trust in the teachings of a credentialed educator. Then what is it? Let's open up (well, actually click on: http://www.newadvent.org) the *New Advent Catholic Encyclopedia*. It defines the act of faith as follows:

> the assent of the intellect to a truth which is beyond its comprehension.

Further, to achieve faith,

> the intellect must assent unhesitatingly, the will must promptly and readily move it to do so.

In this definition, it is the human faculty of will, and not of reason, that plays the crucial role in achieving faith. This position is fully consistent with the rejection by the Catholic Church (as articulated first by Pope Innocent XI in the 17th century) of the notion that faith ultimately rests on an accumulation of probabilities. Far less important, but satisfying, this position is also consistent with our tentative faith math formula in which faith is treated as a separate and distinct entity from reasoned probability. This is the good news for our fledgling mathematical description of faith; however, the bad news may be far more significant. In particular, let's consider what it might mean to view faith in terms of "assent of the intellect to a truth which is beyond its comprehension."

First, I'll remind you that we spent some time in chapter 2 on the notion that a proposition must have a defined meaning before we can attach to it the attribute of being true or false. In the extreme, we could not attach *true* or *false* to a random string of alphanumeric characters. And if neither the value of true nor false can be attached, then we cannot talk meaningfully of the probability that the proposition is true. So if a proposition has no meaning to me (or, indistinguishably, a meaning unknown to me), then this situation would be a nonstarter in terms of expressing my level of confidence, or probability, that it is true. So does the notion of associating faith with a "truth which is beyond [my intellect's] comprehension" serve to sharply and irreversibly distinguish the concepts of reasoned belief and faith? Is the idea that faith and probability are in any way comparable, let alone accessible to side-by-side placement in a mathematical equation, a wild one? Might we just as well try to add together apples and the color orange?

Well, before we send the faith math idea the way of communism, reusable dental floss, and Reaganomics, let's take pause. After all, I wouldn't want a minor technical glitch to impede the birth of a new, fundable discipline: mathematical theology. Even if we tentatively accept the idea that faith, unlike probability, can attach to a proposition with undefined (or unknowable) meaning, this does not preclude the possibility that there *do* exist propositions to which both faith *and* probability can attach. Indeed, the premise of our analysis is that G *is* just such a proposition. I would suggest that there are many propositions to which both forms of belief can be associated. Consider this example:

> Jesus was resurrected from the dead.

While we may lack a scientific basis for explaining the phenomenon of resurrection from death, I think most of us, nevertheless, have a reasonably sound notion of the proposition's meaning. It is that Jesus was biologically dead and then subsequently came to life—not figuratively nor metaphorically, but literally. To this proposition we can surely attach a measure of reasoned probability. This probability, were it to be computed, would be based presumably on consideration of scriptural accounts and supporting historic evidence. But to this proposition surely one can also attach *faith* in its truth. The faith-based belief would be quite different in character from its probabilistic counterpart, having stemmed from a very disparate source. If we are to adopt the understanding of faith suggested by the *New Advent Catholic Encyclopedia*, an element of that disparate source would be the sheer intellectual *will* to believe.

So if Proposition G and the resurrection example are instances of propositions to which both faith and probability can attach, what would be an example of that class of propositions to which only faith has meaningful application—propositions whose very meaning evades comprehension, reflecting a truth that is inherently beyond intellectual grasp? To find such a proposition I took the approach of simply opening the Bible and then reading until I reached an appropriate example. I eventually arrived at

> In the beginning God created the heavens and the earth.
> —Genesis 1:1

Now, perhaps this is a bad example in that it is only *my* comprehension that is challenged. Well, it's much too late in this analysis to attempt concealment of my intellectual foibles, so I'll continue regardless.

What does the proposition mean? Let's begin by assuming that the descriptor "the heavens and the earth" is what in more modern vernacular we would refer to as the physical universe. What does it mean to have created the universe? Since the time of Einstein, we moderns understand that time and space themselves are attributes of, rather than backdrops to, the physical universe. And to give credit where it's due, 15 centuries before the work of Einstein, St. Augustine noted that

> the world was made *with* time and not *in* time.

So if time came into existence *with* the universe, what does it mean to say that the universe was *created*? After all, creation is an act, and an act is something that occurs over a time duration. But there was no *time* available over which to perform the act since time only arrived *with* the universe. You might argue, "But surely we have an intuitive feel for the notion of a *beforeness* to the universe during which the creation was enacted." This view is understandable but is based on the shaky foundation of extrapolating our everyday experience of the physical world to situations in which our intuition fails. After all, how could anyone possibly conceptualize a prespace and pretime setting? In analogy, hypothesizing events prior to the beginning of the universe is like speculating on the existence of an undiscovered continent located north of the north pole.

In my mind the proposition that God created the universe is of a very different nature to the proposition that

Jesus was resurrected from the dead. While I know how to ascribe logical meaning to the latter, I do *not* to the former. Now, it must be understood that to assess a proposition to lack meaning is not to say that it is false. To call it a false proposition is to imply that the negation (i.e., the opposite) of the proposition is true—that is, that God did *not* create the universe. But if a proposition is assessed to lack comprehensible meaning, then its negation also lacks comprehensible meaning. After all, a proposition cannot be made meaningful by simply inserting the word *not*.

In analogy, hypothesizing events prior to the beginning of the universe is like speculating on the existence of an undiscovered continent located north of the north pole.

So why do I understand one of the propositions but not the other? Let's explore this. We use language in general to convey concepts. If I want to express to you the concept that I'm lifting a teapot, well . . . see, I've already done it. I have taken the elemental concepts necessary to construct the aggregate concept of me lifting a teapot, then taken the corresponding elemental words and strung them together into a sentence that represents the aggregate concept. The word elements are the *I*, the *lifting*, and the *teapot*. In this process, concept preceded language. Indeed, it is the concept rather than the representative language that is the object of scrutiny in a Bayesian analysis of truth probability. This is why the meaning of a proposition must be established prior to addressing probabilities. Now, when we are confronted by the revealed word of a scripture, the process must be reversed. That is, we begin with the words and then need to retrofit the source concepts in some way. It is, of course, the case that the inference of concepts from language is not associated exclusively with the reading of scripture but with any literature—scientific, historic, geographic, or otherwise. We are generally able to infer meaning, however, by familiarity with the component words and the context in which they occur. For example, *teapot* can be used in a multitude of settings while maintaining a fairly invariable meaning. I for one tend to use the word habitually. But when the proposition is made that God created the universe, then, clearly, the constituent words are not being used in their established sense. In particular, the word *creation* has well-understood, preestablished meaning to us in terms of an act effected in time, yet this cannot be its sense in the subject proposition.

Can we conclude that it's a metaphor? Nonscriptural literature often relies upon metaphor and simile, particularly in a poetic setting. Can the proposition be anything *but* a metaphor if the literal interpretation of "creation" is inapplicable. A metaphor in general literature conveys a poetic or figurative truth. So can we view faith as something that can be attached to both literal propositions *and* to poetic or spiritual propositions that are intended to convey a sense of truth that transcends flat, logical, mundane truth? Is the truth of the proposition that God created the universe more analogous to the truth that can be ascribed to a Mozart piano sonata or a Turner landscape than the type of truth that can be ascribed to the proposition that Albany is the capital of New York? The bloody climax of Homer's *Odyssey*, in which a priest and a poet beg Odysseus for their lives, is brought to mind. He beheads the priest instantly but spares the poet on the grounds that he dares not kill a man who has been given his divine art by the gods. If we tentatively accept that faith attaches to spiritual truths, what can we infer about faith-belief as applied to a more mundane proposition that *is* accessible to probabilistic analysis? Does this notion of spiritual truth cast any light on the question of how faith complements reasoned probability in assessing an overall level of belief in a proposition to which both can be attached, such as Proposition G?

I am reminded of an observation by the 20th-century classicist Edith Hamilton in her book *The Greek Way* (1993). In contrasting the style of classical Greek poets with that of our more contemporary English counterparts, she wrote that "the English method is to fill the mind with beauty; the Greek method was to set the mind to work." Although her comment was made in the unrelated context of the brilliant economy of Greek verse, I wonder if this contrast has analogy in the dichotomy of faith and reasoned probability. That is, while assessment of probability is something that sets the mind to work, faith is the acceptance of a spiritual beauty or truth that is inaccessible through logic. Let me clarify that I am not suggesting that the truth of holy scriptures lies in their poetic character. Indeed, I do not believe the Bible was intended to be good poetry any more than it was intended to be good science. Rather, I am proposing that spiritual truth, in analogy to poetic, musical, and artistic truths, is something other than the brute truth of logic.

Now, to associate faith with notions of truth beyond logic and with spiritual beauty is to cast it in a very positive light. But is faith inherently good? You'll recall that Chad the mall boy was asked this question, to which he replied,

> I don't think that faith is inherently good or inherently bad. Faith in good things is good, faith in bad things is bad.

He cited the faith of those who perpetrated the September 11 atrocity. St. Thomas Aquinas, who lived in the 13th century, would seem to be a critic of Chad's perspective. He said that

> Faith has the character of virtue, not because of the things it believes, for faith is of things that appear not, but because it adheres to the testimony of one in whom truth is infallibly found.

This perspective would certainly seem to imply that faith is inherently virtuous. St. Augustine, around the turn of the 5th century, said that

> Faith is a virtue by which things that are not seen are believed.

Yet to view faith as necessarily virtuous surely requires that we confine its definition in a way that allows it to associate only with qualifying beliefs. Does this then demand expansion of our taxonomy of belief mechanisms beyond reasoned probability and faith to include a third factor: faith's evil counterpart? By analogy to the New Advent Catholic Encyclopedia's definition of faith, this counterpart would reflect the assent of the intellect to a *falsehood* that is beyond its comprehension.

Thus even in faith we are confronted by relativism. Viewed another way, one person's faith is another's preposterous, arbitrary, and perhaps even satanic belief.

Thus even in faith we are confronted by relativism. Viewed another way, one person's faith is another's preposterous, arbitrary, and perhaps even satanic belief. This relativistic phenomenon is one that can be clearly and disconcertingly observed between religions and even within a single faith. So is this subjective, relativistic aspect of faith something that puts paid to our budding faith math? Is it legitimate to posit that

overall belief = probability factor + faith factor

if there is no agreement even on what can be labeled faith? I'm inclined not to see this as a fatal problem with the formula since, after all, in addressing belief we are considering something that is inherently subjective. Beliefs are subjective, Bayesian probabilities are at least partially subjective, degrees of faith are subjective, and so a little incremental subjectivity as it applies to the nature of faith is not likely to be a silver bullet. We could still insist that for any *one* individual the belief formula incorporating subjective probability and subjective faith would be valid. However, we would need to assent to the notion that the term *faith* in the formula is defined as whatever the subject individual considers to be faith rather than trying to base its definition on a third party or committee finding.

So where does all this lead us in establishing the comparative roles of faith and probability in the formulation of beliefs? Let's write the embryonic formula one more time as it applies to Proposition G:

$$B(G) = P(G) + F(G).$$

In words, the aggregate degree of belief in the existence of God by an individual, denoted B(G), stems from two contributing components: reasoned partial belief, denoted P(G), and the faith factor, denoted F(G). Are we now in a better position to assign and contrast the roles of P and F as they form the sum of belief? Are we now better prepared to answer the question posed earlier of whether it is meaningful to refer to the faith of an atheist?

In my mind, it is a pleasing idea that spiritual truth, the object of faith, is more analogous to the truth of a Mozart symphony than to the truth of the proposition that the population of Michigan exceeds 7 million. In this sense, faith reflects an experiential belief rather than a reasoned belief in a mundane proposition of statistics, logic, or concrete fact. Perhaps probability can be viewed as a snapshot taken in the cold light of logic, whereas faith is more like the exhilaration of experiencing great music. Yet, as we discussed, faith *can* be attached to a proposition with which the more conventional meaning of "true" (or "false") can be associated. Indeed, the premise of our analysis is that G is just such a proposition. In this case, is the idea of experiential truth a meaningful one?

The classical Greek noun *pistis,* as it appears in the New Testament and classical literature, is translated in various contexts to mean "faith," "belief," and sometimes "trust." If we were to replace the word *faith* by *trust* in our deliberations, would it better convey the idea of experiential belief? In this alternative vernacular, we could refer to the trust in the truth of a proposition. Here trust implies an ongoing intellectual reliance on a truth. Perhaps it also implies that there are ongoing consequences of a faith-based belief reflected in the way we live our lives. After all, living is a very experiential sort of thing.

Now, with this experiential view of faith in mind, let's return to the atheists. Can we associate the notion of faith with the beliefs of an atheist, or for that matter with any proposition that lacks religious content? I think that the answer to the latter question is "yes." I know of atheists who, when professing faith in the inherent goodness of people or the virtue of compassion, have as much right to the use of that word as anyone else. It is the assent to such experiential truths that governs their lives. But can someone have faith that Proposition G is false? That is, faith—expressed earlier as $F(G^*)$—that G^* is true? Can there be experiential belief that God does not exist? What would that particular experience be? Why would an atheist be attaching faith to a religious proposition anyway, albeit a negation? I think $F(G^*)$ is void of meaning.

So let's agree that we've sorted this whole thing out. The matter of the reconciliation and integration of faith with reasoned probability has been compellingly resolved. We have, in effect, put future generations of theologians, philosophers, and other profound-thought professionals out of work. But no, this is unconscionable since those future professionals may now be diverted through desperation into careers such as the law, telemarketing, and health care insurance. Let's at least admit to the tentative nature of these conclusions.

I'll finish this chapter by answering a question I posed at the end of chapter 8. I think I now have the math to tackle the question of whether my calculated, reasoned truth probability of Proposition G (of 67 percent) is equal to my actual, overall degree of belief that God exists. My answer is "no." I would set my overall belief factor at

$$B(G) = 95\%.$$

So since my reasoned probability is

$$P(G) = 67\%$$

then use of the faith math formula (subtracting P from B) reveals that

$$F(G) = 28\%.$$

That is, a faith factor of 28 percent is necessary to account for the discrepancy between my reasoned, calculated probability of God and my actual degree of belief in his existence. So this 28 percent factor is my trust in God's existence: the experiential component of my belief. But what if my Bayesian analysis had resulted in a reasoned probability of $P(G) = 97\%$? In this case, the faith factor could be no more than 3 percent since total belief cannot exceed 100 percent. This would severely limit the role of the experiential component of my belief. I think that the math is predicting a reasonable phenomenon here. After all, I imagine it is as difficult to derive spiritual experience from a concrete lump of logical truth as it is from a stone god.

Therefore, I am quite content with my reasoned probability of 67 percent and overall belief of 95 percent. And what of the other 5 percent, the doubt factor by which B(G) falls short of 100 percent? Well, that 5 percent is reassuring evidence that my brain and I are still alive since I suspect the 5 percent will be extinguished once *I* am.

Chapter Conclusions:

Faith is more than an excursion beyond the boundaries of reasoned probability: Faith-based belief is of a distinctive character.

Faith and reasoned probability measure disparate forms of belief, yet they may be viewed as elemental, additive components of overall belief in a given proposition.

Faith-based belief is associated with truths of an experiential nature, analogous to musical and poetic truths, and which can influence the way a life is lived.

There are certain types of propositions to which faith-based belief *only* may attach. These are propositions with exclusively spiritual meaning.

Faith may apply to propositions that are not necessarily of a religious nature since not all spiritual truths are perceived to have religious origin.

I assess my own overall degree of belief in Proposition G to be 95 percent: 67 percent reasoned probability, the balance of 28 percent originating in faith.

eleven

An Existing Question

Question for This Chapter:

What does the word *exists* mean, first as it applies to the material world and then to our proposition that God exists?

This journey is approaching its end. Yet throughout it I have had a gnawing sense of incompleteness: a distant but distinctive feeling of guilt. I won't feign ignorance of its origin. I have certainly done my level best to conduct this investigation by the professional standards of any reputable mathematical theologist—well, more to set such standards, really, in anticipation of the first such reputable analyst. I have established the methodology, I have gratuitously attached the addend *-ology* wherever possible to reflect technical rigor, I have defined all terminology . . . well, almost all terminology, which brings me to the source of the guilt. While key words such as *God, probability,* and *faith* were carefully investigated and their definitions ultimately nailed down, at least for the purposes of this analysis, the one term that received little attention was this one: *exists.* Yes, I seem to have executed this whole study without taking the time to define 50 percent of the words in the central proposition of the analysis:

God exists.

This was the source of the gnawing guilt, the sense of incompleteness. I had convinced myself of the fact that we all surely know what it means to exist. After all, I had promised in the very first chapter that this was not to be one of those impenetrable, scholarly philosophical or theological works but rather a quick, no-nonsense, pragmatic calculation of Proposition G's truth probability. What could be less no-nonsense and less unscholarly than a stultifying treatise on what it means to exist? Thus I had decided to ignore the issue. Yes, the issue *exists,* but so does the well-established analytical concept of *ignoring.* On the other hand, we live in a new age of semantic precision. In this age, prominent figures have asked, for example, what the meaning of *is* is. This enlightenment cannot simply be ignored. So the compromise I reached was to complete the important part of the analysis before I attempted to address the secondary issue of existence. In this way, the analytical waters would not be muddied by the egregiously obscure matter of what it means to exist; yet, at the same time, my guilt would be nicely assuaged.

So what *does* it mean to exist or, for that matter, to fail to exist? As I look around the room in which I sit, none of the items in view—such as furniture, books, window blinds, and slightly overdue video rentals—seem to lack the property of existing. Perhaps I'm in an atypical part of the house. But no, if we define the *universe* as the collection of all things that exist, then we are led tautologically to the conclusion that everything in my study, as well as in the rest of the house, must exist since they are part of the universe. This definition would imply that if God exists, then he is also part of the universe. But some would have it that God is outside the universe—necessarily so if he created it. At this point, you probably share with me a sense of fiddling about meaninglessly with language. I would prefer to leave that sort of professional approach to properly certified philosophers and theologians, so let's start again.

So what does *it mean to exist or, for that matter, to fail to exist?*

Material objects would seem to be an easy starting point at which to consider the matter of existence. We all have a sense of the existence of material objects. We can touch them, see them, and sometimes hear, smell, or taste them.

The last couple of centuries or so have offered much scientific insight into the means by which we achieve these sensings of the material world. Electrons jumping from one energy state to another within the atoms and molecules that form a material object are obliging enough to create electromagnetic disturbances that travel rapidly away from their source, meeting possibly with an observer who is then able to confirm the object's existence. Certain material objects, such as a played piccolo, create rapid air vibrations. These vibrations are the elements of a sound wave, which again can make its way to an observer (listener)—albeit at a less alarmingly hurried pace than that of an electromagnetic wave, although it nevertheless arrives eventually. Then the refusal of certain types of fundamental particles to occupy the same space as their own kind, characterized in the 1920s by the German physicist Wolfgang Pauli as the exclusion principle, is what ultimately creates the back-pressure we experience when touching a solid object. The degree of heat associated with touch is dependent on how rapidly the constituent molecules of the touched material are vibrating. The vaporization of molecules from a material object and the migration of those vapors to an observer are the means by which we can confirm the existence of odorous or tasty objects. Then at the back end of all these existence-confirming processes are the mechanics and electrochemistry of our biological systems, which sense the incoming signals, route them, and ultimately produce the appropriate neuronal sparks between the ears, at which point the perception of the object is a done deal.

Still, we probably did not need all that scientific understanding to know intuitively that any given material object—be it a chair, a bacon sandwich, or Lake Huron—exists. But then, it must be admitted that science has considerably widened the range of those things that we consider to be material. For example, the gases that constitute the air around us had always been somewhat inconspicuous. Even though as early as the 5th century B.C. Empedocles recognized air as one of the four elements, it was not until the 17th century that the material nature of air as a composition of gases was understood. Recognition of the materialness of light itself goes back to the 17th century when Isaac Newton made an early attempt to describe light as a composite of particles, although it was not until the early 20th century and the birth of quantum theory that the material, particle-like nature of light began to be understood. In the world of modern physics, we are now aware of a menagerie of fundamental and composite particles that can account for all the types of matter we have thus far detected. Indeed, the very forces by which matter exerts influence on other matter are now understood to be mediated by yet other forms of matter. (I use the term *matter* in a general sense to mean something that is composed of material objects.) So photons are the media of the electromagnetic force, W and Z particles of the weak nuclear force, gluons of the strong nuclear force, and gravitons of the gravitational force.

All these particles are the stuff of matter and thus of the physical universe. Even space–time itself—which has been viewed traditionally as the arena in which matter exists, the stage on which it plays—is now being viewed by physicists interested in the first instant of the universe as itself comprising discrete quanta of space–time stuff.

So all this matter in aggregate is the physical universe. But is this the whole universe—that universe we define as the collection of all things that exist? Well, before we consider the subtle question of what exists in addition to matter, let's first address some of the subtleties of matter itself. The subtleties of which I speak are those that are thrust upon us by quantum theory. While this theory has been an enormously successful one by any standard in terms of its explanation and prediction of physical phenomena, it has nevertheless painted for us a rather strange picture of matter as it exists at the microscopic level of fundamental particles. Take the photon, for example, the quantum packet of light or any other form of electromagnetic energy. Photons exist—but what are they? Well, whatever a photon is, it sometimes behaves like a little particle, a sort of tiny billiard ball. In fact, it was Einstein himself who first suggested this particle-like characteristic of light (at least in the modern era) to explain something known as the photoelectric effect in which certain materials when irradiated with light will emit electrons (another species of particle). He deduced that there must exist discrete little lumps of electromagnetic energy to account for the unexpected energy characteristics of the emitted electrons. So far, so good. But here's the strange part. A photon can also behave like a wave, sort of like the ripples on the surface of a pond. For example, light can form interference patterns like intersecting water ripples or can be polarized, as anyone who dons cool sunglasses knows. Indeed, before the advent of quantum theory, electromagnetic phenomena were generally understood in terms of wave principles.

Well, it turns out that photons are not unique in having these quirky, ambiguous characteristics. Electrons, protons, neutrons, and all the exotic zoology of particles are really part particle, part wave. So how can something be both particle and wave? Our intuition, based on the large-scale world of our everyday lives, is that an object must be made to choose: It cannot be both. Perhaps a better way of considering these tiny, confused objects is to view them as neither particle nor wave. Some have suggested that we call them "wavicles" to reflect the fact that they are not particles and not waves but are simply what they are: something else. While exhibiting features of both waves and particles, these wavicles are in fact their own thing.

To expect microscale systems to conform to our macroscale intuition is one of the worst forms of bigotry and political incorrectness: It is scalism.

If these existing entities are neither wave nor particle, then why do we insist on making these comparisons rather than just accepting them as something else? The reason is that we humans have our roots in the macroscopic world. We are lumbering macro-yobs who need to make analogies with large-scale phenomena; otherwise, we simply cannot conceptualize these tiny mysteries. We humans are generally not too good at conceptualizing things that are not . . . well, that are not billiard balls or water ripples or other everyday macroscale phenomena. To expect microscale systems to conform to our macroscale intuition is one of the worst forms of bigotry and political incorrectness: It is scalism. We simply want all phenomena at all scales of existence to behave in the same way as the intuitively sensible macroworld in which we live. Well, it turns out that the microworld of quantum systems simply does not comply with this macrocentric expectation. To exist at the quantum level is a rather different form of existence.

For example, we macro-yobs expect an object to be describable in terms of its precise location and speed. As I drive my car north along Interstate 71 to Cleveland, I will determine at some instant that my position is at the I76 exit for Akron and my speed is 63 miles per hour. There's no impediment in the macroworld of interstate driving to having simultaneous and relatively exact knowledge of position and speed. Now, electrons, to pick a fundamental particle at random, behave nothing like cars. In their microworld of existence, they know nothing of interstates and Cleveland. Even the notions of position and speed are ones with which they are not entirely comfortable: concepts imposed upon them by our parochial, macroworld prejudices. What's more, these electrons are stubborn and bloody-minded little bounders. They have a certain pride in their own form of existence and are determined not to let the macro-yobs have it all their own way. This is how they stand up for themselves. In its natural state of existence, an electron can best be conceptualized in macroterms as a set of probability waves that define the likelihoods that the electron has certain properties, such as a specific position, a specific speed, a specific orientation, et cetera. So instead of the electron having a single position, as we might expect, it is really somehow smeared over a range of possible locations. Say one of us macrotypes then comes along and demands to know the exact position of the electron and, what's more, has available to her the clever measurement techniques to establish that position. Well, by making that measurement, she actually affects the probability wave that applies to the electron's position, forcing it to shrink in on itself to a single point such that the electron's position can now be specified accurately. But the way in which the peevish, oppressed electron retaliates is by spreading out the probability wave that defines the possible speeds of the electron. (More strictly, what becomes smeared is the momentum of the electron, which, if the electron is moving at significantly less than the speed of light, is the electron's speed multiplied by its mass—but I'll take the liberty of focusing on the question of speed.) So the macroyob may have won the battle by nailing down the electron's position, but the price she paid is now knowing nothing of the electron's speed. This aggravating attribute of quantum systems was mathematically characterized by the German physicist Werner Heisenberg in the 1920s and is referred to as the uncertainty principle.

Similarly, had the macro-yob successfully nailed down the speed of the electron, then she would suddenly have known nothing of the electron's position. Accuracy in the knowledge of one is always at the expense of accuracy in the definition of the other. This behavior is, or should be, strange to us. If we insist on conceptualizing this aspect of quantum systems, then we might envision the very act of observing the system as being implemented by the slamming of photons into it. Thus observation itself has the effect of shaking up the very system that is being observed. This does not happen, or at least does not happen noticeably, with macroscale systems. Observing that my car is traveling at exactly 63 miles per hour does not have the effect of catapulting the car instantly to parts unknown.

So the existence of matter at the microlevel is a very different type of existence from that of matter at our everyday macrolevel. Yet even though we cannot fully conceptualize quantum systems, we do not doubt their existence—indeed, macrosystems are the composites of microsystems. So we accept material existence even when it cannot be conceptualized. Now to be fair to physicists, we *do* understand these strange aspects of quantum systems when expressed in the language of mathematics. Indeed, as mentioned previously, quantum theory has provided a staggeringly successful picture of the physical world. It is only when we make the attempt to form analogies with the properties of macrosystems that strangeness becomes rife. (By the way, it is a source of some relief to me that those who view the random, probabilistic aspects of biological natural selection as a threat to their religious beliefs are not equally concerned by quantum theory and its profoundly probabilistic implications. Will the theory of intelligent design one day be extended to include the hypothesis of the intellectual electron?)

Anyway, this collection of matter and the arena of space–time in which it performs is what we consider the physical universe. But there are surely things that exist that are something other than matter—a thought, for example: a clever idea or an emotion such as aggravation. When we have a thought, could it be argued that that thought does not really exist? I think not. I suppose someone who is terminally mechanistic might wish to claim that a thought is no more than an electrochemical process in the brain and so might be understood (or is at least potentially understandable) in terms of the properties and organization of matter. However, I would think that the language of electrochemistry is an unuseful one in which to define thought. Recasting analogies introduced earlier, this would be like characterizing the plot of a television drama in terms of a sequence of electron emissions from the filament of a cathode ray tube or a violin concerto in terms of a time-dependent air pressure distribution. Well, I suppose that even if one concedes that a thought may be viewed in terms of complex patterns of material behavior, we surely acknowledge a significant difference between the notion of existence as it applies to a material object, such as a chair, and to a thought. In particular, a thought exists purely between our ears.

Now, it is the case that a chair only becomes appreciated as such once a sequence of events has been completed that begins, say, with the emission of light by the chair and culminates with an electrochemical process between the ears. However, what differentiates appreciation of the chair's existence from, say, a good idea is that the former is directly attributable to an external stimulus, whereas the latter is not. Now, René Descartes, in the infinitely pragmatic spirit of French thought, pointed out in the 17th century that "I think, therefore I am" is the only thing of which he could be sure. So who knows if there *is* really any such thing as an external stimulus: Perhaps the whole universe is no more than a mind game. Let's resolve to avoid that perspective since I'm disinclined to complete a book for an audience that I have only imagined. (Not that this criterion dissuades *all* authors.) Instead, let's adopt a tentative classification of existing things into those that rely completely on our brains for their existence and those that have independent, exterior existence. But is this necessarily a robust classification scheme?

When we have a thought, could it be argued that that thought does not really exist? I think not.

The ancients had some interesting ideas about the relationship and distinction between the material world and other forms of existence. Plato had his theory of forms. These forms were idealized objects of which the material world could produce only crude facsimiles—templates that are imperfectly applied in the physical universe. For example, a perfect circle would be considered something that exists as a form. Plato argued that even if the material world never successfully produces a perfect circle, we can surely state nevertheless that a perfect circle is an existent entity. Here, Plato appeared to view forms as something that are objective and exist "out there"—that is, entities that do not rely purely on inter-ear processes for their being.

St. Thomas Aquinas in the 13th century proposed another interesting dualistic description of the world by introducing the language of essence and nature. Whereas the *essence* of an entity is that set of characteristics that makes it what it is, its *nature* is the way in which it operates in the world and makes itself known. Here, a duality is proposed that distinguishes the inner character of something from the way it performs in the material universe. These ancient and medieval ideas about the nature of things might appear rather obscure to us moderns and perhaps even a little naive in the face of pragmatic, modern science. But then perhaps not. For instance, in what sense does the Beatles tune "Yesterday" exist? It exists most substantially for me when I am listening to it or humming it. I suppose it exists also when at least someone, somewhere is listening to it or humming it. But what about the unlikely circumstance that at some single point in time it is not being given radio or television airtime; it is not being played on CD, record, or tape; and nobody is humming or whistling it? In what sense does "Yesterday" exist at that specific point in time? Well, perhaps we could argue that the song's form or essence still exists even when unplayed. If this is so, did its form always exist from the beginning of time the way the form of a perfect circle always exists? Did it take Sir Paul McCartney to channel or capture that form such that it became realizable in the physical world as a musical experience? Was that morning when McCartney awoke with the melody in his head the first time at which the form of "Yesterday" became mirrored by something in the material world? More practically, are forms copyrightable, thus allowing me to establish ownership of future compositions by others? Of course, I do not have any answers to these questions but feel confident in concluding that the notion of existence is one that applies to a much broader range of entities than material objects and the structures they form.

Now when we talk of God's existence, how does this concept fit into the crude taxonomy we have identified? Well, first of all, perhaps we can dismiss the idea that God is of a material nature in the sense of being composed of common matter, like a chair. Belief in a man residing on a cloud or a mountaintop is surely not among mainstream views. But can we say that God is like a chair at least in the sense that he has objective, external existence and does not reside exclusively in our thoughts? Well, the God of our analysis—the God of the monotheistic faiths—is certainly believed by those of faith to be more than an idea. So indeed, when the word *exists* is employed in Proposition G, it is intended to reflect an objective, external form of existence.

Imagine that everyone believed in God's existence but was wrong—what a magnificent error that would be.

As an aside, imagine that everyone believed in God's existence but was wrong—what a magnificent error that would be. Continuing with this irrelevance, if you had the following two options, which would you choose?

Option 1: To live in a universe in which God exists but where few take much notice of him.

Option 2: To live in a godless universe in which everyone erroneously believes that he exists and acts accordingly.

I wonder which option God himself would select for us had he the choice. But, then, if God is the very reason for recognition of the good, as discussed in chapter 6, then Option 2 is a fallacy since the good would be undefined.

Well, back on track: If God exists, it is in an objective but nonmaterial fashion. I suspect that additional profundity of thought could be shortcut by jumping to the conclusion that the way in which God exists will have no analogy in the existence of other entities. But if his existence has no analogy to other forms of existence, we might ask on what basis the word *exists* is adopted in application to God? Well, let me proceed to explain why I would consider this to be a rather arrogant question, were it to be asked. Reconsider the quantum-scale system of a single electron. Using the language of mathematics, we have developed a vernacular in which we understand the relationships between the measurements we make of that system, such as those of position, energy, and momentum. We can even introduce words (such as *wavicle*) that are shorthand for the general way in which these measurements behave, yet what do we know of the entity that is behind the measurements? To adopt a medieval term, what is the *essence* of the electron—that cause behind its physical interactions with our measuring apparatus?

This vagueness about the essence of the quantum world then brings home our prior naivete about our understanding of the macroworld, which is, after all, simply an aggregate of quantum systems. Again, there is much we can measure about a chair using the evidence provided by a tape measure, a weight scale, and other instruments. But what is the essence behind the weight, the height, the color, and the surface texture? What is that thing producing the measurable characteristics that we have chosen to model and understand in terms of phenomena like atoms and electromagnetism? Scientists, at least while wearing their science hats, might scoff at such a question since the domain of science is the measurable. Yet after the scoffing, the question is left unanswered.

God, like the chair, provides evidence of existence. Our perception of God, like our perception of the chair, is based on the sum of evidence to which we are exposed. Now it must be conceded that the evidence for the chair is of a somewhat different nature than is the evidence for God. The former entails the detection of electromagnetic radiation and of back-pressure from touch, while the latter, in my opinion, takes the forms of evidence identified in chapters 6 and 7. If there is any significant disparity, it is that science provides a framework in which to make predictions about the ways in which certain types of evidence, or measurements, will relate to one another. For instance, if I measure the details of a chair's dimensions and measure the average density of the timber from which it was constructed, then I can make a prediction about measurement of the chair's weight. Thus scientific analysis has allowed me to relate three sets of measurements, and in so doing I might convince myself that I know something of a chair. In contrast, the evidence for God cannot be modeled and interrelated in this way. We lack that type of understanding when it comes to God. But when all is said and done, we know both God and the chair only by the evidence that's revealed. So what *is* the essence of God behind all the evidence? In what sense does he exist? Well, how can we expect to address those questions about God when we cannot address them even for furniture?

I suspect that I have shed no light for you on the notion of existence as it applies to God, but I hope I have at least cast some shadow on the notion as it applies to the mundane. I would conclude that the mystery of existence is not uniquely associated with the divine. So whatever was your original intuition about the meaning of *existence*—stick with it, as it's probably as good as any other.

Is there hope that one day we will fully understand the nature of God's existence? Perhaps that's a treat in store for us in the afterlife—where form, essence, and the inner realities of things may be the stuff of direct perception. Though again, perhaps even *then* we will only get to see a few additional facets of God. That question will continue to exist.

Chapter Conclusions:

As we have come to understand more of the physical universe and its quantum nature, our comprehension of the underlying essence of material things has receded.

That God's nature is obscure to us should be no surprise when even the mundane poses a mystery.

---

twelve

Faith AfterMath

Questions for This Chapter:

What are the collateral benefits of quantitative analysis?

Can an individual's faith and reasoned probability influence one another?

Does our model of faith and probability cast any light on the nature of religious conflict?

Can faith be a source of evil?

Going through my monthly bank statement can be a miserable business. The way such small individual numbers conspire to produce such tremendously large sums is a mystery of arithmetic next to which the 23 unsolved mathematical problems posed in 1900 by David Hilbert pale. Still, I do it every month, and not so much to confirm the bank's math but more to sit back and get a feel for where the money is going. So as I'm doing the arithmetic, the real insight is coming from my incredulity at having accounted for eight pizza checks or my astonishment at the price of a haircut. In this sense, many forms of quantitative analysis are like checking a bank statement. Although we are interested in the numbers ultimately produced, as much if not more insight comes from the detailed thought process into which we are forced in order to complete the analysis. Issues are opened to us that we might never had broached in a more cursory type of evaluation.

For much of my career I have been involved in assessing the risk of a major accident posed by the operation of complex industrial facilities such as nuclear power plants and chemical process facilities. I can vouch for the fact that there is no better and more efficient way to become gruelingly familiar with the design and operational intricacies of such facilities than by trying to figure out what could go wrong and with what probability. From my perspective, the probabilistic analysis of Proposition G was no exception to this rule. While the objective of producing an ultimate truth probability was what provided the tracks along which the analytical train chugged, it was the view—the systematic consideration of the underlying issues—that I found to be the greatest reward.

Still, it's nice also to have a number to show for the effort. What's more, I find the number enlightening—not that it's 67 percent rather than, say, 59 percent but more its ballpark value. In particular, this probability number, produced as my best attempt to analyze evidence, is unequal to the degree of belief in God that I would have assigned directly, in the absence of the analysis. I calculated 67 percent but assess my overall degree of belief in the truth of Proposition G to be 95 percent. In chapter 10, I identified that discrepancy with the effect of faith. In this view, faith is a component of overall belief that acts to supplement that element reflecting reasoned probability. Let's finish by considering a little further the implications of this perspective on faith and its fearlessly proposed mathematical description. For example, while the contrast between faith and probability was explored in chapter 10, what of the synergies between the two: How do they interact and thus affect one another? And furthermore, can the faith math shed light on the nature of religious conflict or, more specifically, on conflicting theological views?

First, let's reconsider whether there really is a bright line that separates the two forms of belief, a line that would justify my mathematical picture. I suspect someone could argue that my faith actually played a surreptitious role in my analysis of reasoned probability. Therefore, to neatly isolate the faith factor in the belief equation $B = P + F$ is illegitimate. That is, it could be claimed that the calculated value of P itself was dependent on the way in which my faith insinuated itself into my *reasoned* analysis, and so to claim that the role of faith-belief can be captured entirely by the solitary mathematical term F is questionable.

Maybe this would be a fair observation since I would not know how to entirely suppress the effect of my faith in the calculation of probability. Yet one cannot avoid the conclusion that a discrepancy ultimately existed between my reasoned probability and my overall level of belief in Proposition G. This discrepancy I would argue is due to the overt role of faith, even though there may have been some faith seepage into the calculation of probability itself.

What might be a more compelling proof of the independent roles of faith and probability? Can we envision a situation in which an individual assesses the absence of one but the presence of the other? Hypothesize someone who assesses the reasoned probability of God to be vanishingly small and yet maintains a strong faith. This seems an unlikely circumstance. For example, while an atheist would meet the first of these criteria, she would certainly not meet the second. Similarly, it is difficult to envision a person of faith who would assess the reasoned probability of God to be very small. This observation would seem to imply a significant positive correlation between faith and probability. That is, the greater is one, the greater the other is likely to be. Therefore, an argument for true independence of faith from probability, and vice versa, might be a difficult one to defend. Then what of the reverse situation? Could we envision a person who has established a large reasoned probability of God but then holds little faith? Indeed, a large reasoned probability would leave little role for faith, as was discussed in chapter 10. There are those who claim logical certainty in God, but one must be somewhat skeptical about their conception of logic.

Then what of the relationship between faith and probability as they perhaps vary over the lifetime of a single individual? As one fluctuates with the ups and downs of life, what of the other? I have not been reticent thus far to stress the subjective nature of probability, whether it applies to God's existence or to the medical diagnosis of an itch. Since the calculation of P(G) is linked strongly to one's life experiences, then any one individual might produce varying truth probabilities throughout his life. What might a lifetime graphical plot of P(G) look like? Perhaps it would tend to be a flat line with the occasional blip corresponding to an extreme life event. That is, his probability might be relatively stable, yet now and again his experiences, both good and bad, tweak that probability, although it eventually settles back to an ambient level characteristic of that person. Then again, perhaps for some the lifetime probability graph would look more like the price action of pork belly futures: up and down continually.

How might faith respond to such probability action? Is it a stabilizer—that is, a dampener that compensates for probability fluctuations thus keeping overall belief at a steady level? Or is it an amplifier that magnifies probability swings? Or does it depend on who you are and the events you experience? I can't answer those questions for anyone else, but for me faith is a dampener that picks up the slack when reasoned probability recedes. But then I am fortunate that my life circumstances have never given cause for a large probability recession. The lives of others have been less fortunate. In her book *A History of God* (1994), Karen Armstrong relates events that occurred in the Auschwitz concentration camp in which a group of Jewish prisoners put God on trial.

> They charged him with cruelty and betrayal. Like Job, they found no consolation in the usual answers to the problem of evil and suffering in the midst of this current obscenity. They could find no excuse for God, no extenuating circumstances, so they found him guilty, and, presumably, worthy of death. The Rabbi pronounced the verdict. Then he looked up and said that the trial was over: it was time for the evening prayer.

Few of us have known such suffering. We addressed abstractly in our analysis the reconciliation of evil with God's existence, it being a factor that ultimately played against the truth of Proposition G, yet I cannot imagine the perspectives of those who are amidst the most acute evils. Surely in these circumstances one's reasoned probability of God's existence, or at least the existence of the good God of our faiths, could be very low. Still, this story ends with evening prayer. Was it faith that was called upon to bridge the chasm formed as reasoned probability receded? Then stories are related in which life's events appear to shatter faith, at least temporarily.

I must concede that to view faith and probability as two elements uninfluenced by one another, as independent aspects of belief, is much too simplistic. Yet while not independent in that sense, they are in my mind distinctive nevertheless: one a reflection of logical consideration, the other of a more transcendent, experiential nature.

In positing this distinction between faith and probability, I suggested in chapter 10 that there is a certain class of propositions to which faith can attach but that reasoned probability cannot. These are propositions that foil attempts at analytical resolution into component, logical concepts since the words from which they are composed are clearly not intended to convey their common meanings. That is, the author of such a proposition had a spiritual sense to convey but lacking good tools to do so borrowed common language despite its ill design for the purpose. It is as if Michelangelo had borrowed a can of spray paint. Nevertheless, the writer took his best shot and hoped that those who read his words would derive from them the sensations that went into their formulation. To

The author of such a proposition had a spiritual sense to convey but lacking good tools to do so borrowed common language despite its ill design for the purpose. It is as if Michelangelo had borrowed a can of spray paint.

such propositions, mundane logical truth and probability have little relevance, and it is faith alone that forms the basis for belief. Perhaps music provides an analogy. We can document music on manuscript, but that series of dots, circles, and lines has no intrinsic musicality. Now when those flat, dull symbols are properly interpreted, something thrilling can occur. Similarly, when I am told that God created the universe, the logical, mechanical notion of creation could not be further from my mind. Rather, it is a sense that purpose and compassion are imbued in the world, and . . . well, there I go using common language.

What might be some of the implications of this notion that belief in the truth of certain propositions can be faith based only? Well, since conflicting propositions are the stuff of intellectual exchange, debate, and sometimes even confrontation, we might ask how faith-based propositions relate to religious conflict. With such commonality in core beliefs between and within the major monotheistic faiths, it is to many of us breathtakingly absurd that differing theologies can be the source of tensions, debate, and much worse. Yet they are. Let's first establish some principles by considering a mundane, nonreligious proposition:

> Albany is the capital of New York.

Imagine two individuals with conflicting viewpoints on the truth of this proposition. The first says, "I'm almost certain that Albany *is* the capital." So individual #1 attaches a high probability to the truth of this proposition. However, individual #2, the product perhaps of a more progressive educational system, says, "I think it very unlikely that Albany is the capital." So individual #2 attaches a very low probability to the truth of the subject proposition. Now, we can apply the math of probability theory to conclude that individual #2 must attach a high probability to the proposition that is the negation of the subject proposition, in other words, that

Albany is *not* the capital of New York.

Hence there is overt conflict. Individual #1 attaches a high probability to one proposition, whereas individual #2 attaches a high probability to that proposition's negation.

Now let's consider the analogy of this conflict where the beliefs are represented not by probability but by faith. Assume the subject is one of those propositions to which I have suggested faith, but not probability, can attach. For example, I believe that

Jesus is the son of God.

To this proposition I attach my faith. It is a belief around which I experience many aspects of my life and make certain kinds of decisions. Now a follower of a different body of religious beliefs may not attach his faith to the truth of this proposition. Yet this is surely not equivalent to him attaching faith to the negation of the proposition,

Jesus is *not* the son of God.

That is, the math of probability does not transfer to the notion of faith-belief. Otherwise, we would need to assume that the negation of the proposition, like the proposition itself, can be assigned spiritual meaning. I cannot imagine what meaning and experience can be derived from the negation in this case. So there is surely no basis for conflict. Indeed, that follower of another faith will have his own set of spiritual propositions, some of which would be outside my own faith-belief, yet I do not attach faith to the negation of his beliefs. Faith is simply not available for that type of application. So I find it difficult to imagine true spiritual conflict.

With such commonality in core beliefs between and within the major monotheistic faiths, it is to many of us breathtakingly absurd that differing theologies can be the source of tensions, debate, and much worse.

Then again, political and even military conflicts in which the participants can be identified by religion are, tragically, numerous. Of course, it is often the case that national, ethnic, or tribal interests are the true sources of conflict, religion being little more than a convenient descriptor of the opposing sides. Also rife is political debate in which religious thought forms the basis for at least one of the conflicting positions. Again, the source of this conflict generally stems not from the validity of personal religious belief but from the insistence that that belief be shared or that its implications be imposed upon society. So although faith and probability are concepts with sufficient commonalities to allow their collaboration in the development of rounded belief, it is their differences that undermine any legitimate basis for faith-based conflict.

You may feel that none of this business of conflicting religious beliefs and the role of faith could possibly be rendered comprehensible without the perspectives of The Axe and Chad. That so many philosophers and other great thinkers have been able to convey their ideas without recourse to mall-based vignettes is testament to their genius.

LATER AT THE SHOPPING MALL: A FINAL VIGNETTE

Background: Thirty years have passed since the anthropic epiphany at the mall; CHAD*'s infections have long since healed. After earning a master's degree in business administration from the prestigious Chuck University,* CHAD *rapidly ascended the corporate ladder to become chief financial officer of one of the largest corporations in the Midwest. Like many in that position during the early 21st century, he did jail time, although significantly less than the average for a* CFO*, for which he was lauded by his peers and much* *sought after for postluncheon speeches. He now works occasionally as a highly paid consultant to the big two accounting firms, but otherwise he considers himself to be in early retirement. He lives in an affluent suburb with his aging mother.*

After false starts in financial consulting and insurance sales, ANAXAGORAS *(*THE AXE*) became a man of the cloth. This had been his calling. For most of his ministerial career, he has served a small rural community, working out of a red-brick church surrounded by fields of rippling corn.*

Scene: CHAD *is sitting on a bench by a clump of plastic flora in a more secluded thoroughfare of the mall.* THE AXE *happens by.*

THE AXE: Chad? Is that you?

CHAD: Can't you people just leave me alone, you media bloodsuckers?

THE AXE: It's me. The Axe!

CHAD: Axe?

THE AXE: Yes.

CHAD: Axe. I can't believe it.

[THE AXE *sits by* CHAD.]

THE AXE: How many years has it been? Must be at least 20.

CHAD: At least.

THE AXE: You know, this is exactly where you used to come and hide out after one of those big blow-out arguments with your mom.

CHAD: [*Awkwardly.*] So, 20 years.

THE AXE: At least.

CHAD: But you know what, Axe . . . I feel like I've been keeping in touch with you just reading those books of yours.

THE AXE: Oh, you've read some of them?

CHAD: Every one. *The Calculus of God, God by the Numbers, God: It All Adds Up, The Divine Abacus, God Has Your Number,* and then the more quantitative works—read them all! And don't think I didn't see some of our old conversations hidden in there, you sly dog.

THE AXE: Hope you didn't mind.

CHAD: Heck no. But there are a couple of things I've been waiting a long time to take up with you.

THE AXE: Really?

CHAD: Yes, especially this business of spiritual truths and why theological conflict is generally stupid.

THE AXE: I don't think I said that . . . stupid.

CHAD: Jackass!

THE AXE: No, no—not you stupid . . . you said . . .

CHAD: Oh, oh, sorry. Flashback there. It's great how we can just pick up after all these years.

THE AXE: I suppose so.

CHAD: Anyway, here's what I think you've been saying. Belief in certain propositions can only be attributed to faith-belief and not to conventional probability-type belief. Now, for these propositions, it doesn't really mean much to talk about their negations, their opposites. That is, if you put in the word *not* to produce a negation, as you would in a common proposition, then you wind up with a proposition in which it's meaningless to have faith, such as "God did *not* create the universe," "Jesus is *not* the Son of God," and so forth. So if there's no basis for faith-belief and none for reasoned probability, then there's no basis for *any* form of belief in the negation. So . . . no conflict!

THE AXE: Yes, that's more or less it.

CHAD: But that's a specious endorsement of faith if I ever heard one. All you're saying is that if a proposition has no meaning, then it can't be assessed wrong. In other words, these propositions don't even meet the threshold at which they earn the right to be true or false. Isn't that . . . well, a bit condescending?

THE AXE: As usual, my old pal, you've only half got it.

CHAD: [*Regressing.*] Oh, I see. Well, explain the other half, the other 50 percent, the oh-point-five that I've missed, Mr. Minister o' Math.

THE AXE: All I mean is this: There can be truths that transcend simple logic. I suppose you could say that they transcend the capacity of the intellect to comprehend, although that's not really it since it isn't the intellect to which they appeal. These are truths that, for want of a better word, could be called spiritual. They are the kinds of truths that affect the way you decide, what you see in the faces of others, and that give it all . . . well, a sense of purpose. It's difficult to convey these ideas. I might just as well try to explain Beethoven's Ninth. Sure I could count the movements; identify the instruments; describe the melodic and harmonic structures in terms of intervals, values, and chords; talk of the instrumental color, the emotion and the rhythm, but would you then know Beethoven's Ninth? My description would only be *about* music and not *of* music. I think that descriptions of spiritual truths are equally inadequate. But if you believe in these truths, then faith is the mode of that belief. And that's a kind of belief that walks with you—that colors your experience of whatever comes round the corner.

CHAD: Do you think that faith-belief is automatically associated with the good? Can it ever be the cause of decisions and acts that might be considered evil, that cause human harm?

THE AXE: I don't know. On the face of it, people do seem to do evil based on their religious views. As Blaise Pascal once said, "Men never do evil so completely and cheerfully as when they do it from religious conviction." Sadly, Pascal's view has stood the test of time. Yet I wonder if it's really faith that drives people to evil acts, or if it's other motivations for which religion is a convenient means of justification or interpretation. I suspect that those who commit abominable acts in the name of God have, in fact, identical motivations to their secular colleagues in evil: political gains, perhaps sheer desperation of circumstance, fear of change, or for some, sadly, a true delight in doing harm.

CHAD: But is faith-belief something that's necessarily attached to the divine, to God? It seems that your mathematical formulation and description of faith might not be confined to religious faith alone.

THE AXE: That type of faith is the one I think I understand most: faith that there's purpose and that there's value in compassion somehow inherent in the nature of things—woven into the fabric of the world. Yet I know of those whose faith in the good seems to lack religious underpinning, or at least they *believe* it does.

CHAD: I'd like to believe that faith itself is God-given. That way we *know* that beliefs resulting in human harm could not be the product of true faith.

THE AXE: I can see how that would be reassuring. But if it's true, then why would faith be withheld from some? Yet again, perhaps it isn't withheld . . . if we want it, then we can have it . . . for if it's thrust upon us, wouldn't that compromise our free will? God *does* seem to value free will above many things: above the prevention of evil and perhaps also above the experience of faith.

CHAD: Why does he so value free will?

THE AXE: Don't we *all* value free will of the individual? Look at the oppressed societies that lack it.

CHAD: But our society values it because we're all aware that we can't and shouldn't trust others to know what's best for us.

THE AXE: And God surely *does* know what's best for us . . . so why should he value free will? Is that where you're going?

CHAD: Yes.

THE AXE: But without free will, how can we choose the good? Isn't good only really good when there exists the option of evil? How could we be meaningfully responsible for our welfare and that of others if we lack options?

CHAD: I see the logic. But couldn't God have created a different logic? A win-win logic?

THE AXE: I don't know, Chad. I don't know if logic is a given, something that even God has to work around. Some seem to think so.

CHAD: And besides, what's the good inherent in this whole goodness-testing experiment that God has set up? Might not *nothingness* have been a better option overall if the grand measure of the good is the balance of world despair against joy?

THE AXE: The logic does seem to rely heavily on an afterlife, doesn't it? Or perhaps net human happiness *isn't* the right metric. I don't know the answers, Chad . . . all I can advise is that you avoid anyone who says they do. We can't hope to see the big picture, but that doesn't affect my faith in knowing the right thing to do today. That's all part of Faith with a big "F"—knowing there is a plan and that it's a good one.

CHAD: You know, I'm going to rent myself a horse and buggy and come out to listen to one of your sermons.

THE AXE: I'd like that, Chad, but don't forget your calculator.

CHAD: I won't.

Curtain falls.

Postscript: ANAXAGORAS *went on to become the recipient of the first Nobel Prize awarded for mathematical theology for his groundbreaking work* The Religious Demography of Heaven: Likely a Blow for Many. *The renewed relationship between these friends inspired* CHAD *to come out of early retirement. He took a teaching post at Chuck U. and eventually became dean of the business school.* CHAD *and* THE AXE *have a standing appointment to meet quarterly by the clump of plastic flora, where they eat heart-plugging food and discuss the infinite.*

Thus the journey ends. So let me now return briefly to one of the principal reasons I noted for taking this short but difficult trip. I said I had wanted to explore a concept of God to which I had been starkly reintroduced upon arriving in the United States: the person-God—the very concrete, substantial God of the monotheistic faiths with whom I may have lost touch a little over the years. It seems that at the end of the day, my analysis has revealed a high degree of belief in his existence. Now in reaching this conclusion I have largely tried to avoid areas of religious disagreement about the specifics of our person-God (although not always successfully, you might observe), focusing instead on the commonalities in belief across and within the monotheistic faiths. To be sure, there are different points of view, and at the risk of aggravating you at this late stage, let me offer a few of my own.

The unprecedented freedoms and energy of this nation assure us a certain scale and breadth in all things. The American spectrum captures the very best and the very worst our freedoms allow. Religion is no exception. And these extremes I measure against the most important of metrics: the degree of compassion that religion imbues. Sure, there are other superficial measures of religious diversity. For example, like many of overseas origin, I was incredulous and ultimately saddened at the existence of religion-inspired debate on physical and biological origins. Here I saw the danger that some might equate the truth probability of the Christian message with the truth probability of a literal reading of biblical creation. Whether or not this is a real danger, this particular debate is one ultimately of the intellect and of science, and the winner must be the side with the better science. I think it a debate irrelevant to God's compassion and purpose. And I have come to see through experience that the side of this debate on which any one individual sits is no better an indicator of his moral behavior than is his toothbrush color.

The American spectrum captures the very best and the very worst our freedoms allow. Religion is no exception.

While we're in this general area, you will have noticed that I have taken a few cheap shots at the theory of intelligent design—a rejection of the mechanism of natural selection as the route to the current diversity of life. I can see how intelligent design would be an attractive idea, more consistent with an understanding of a God-created world. Indeed, our comprehension of how our world stems from God would then be nicely framed: He designed it as would an engineer, and then he built it. But consider the arrogance, or perhaps naivete, implicit in this reasoning—that we should have such a firm handle on the workings of God. A recurrent observation in this analytical journey has been the lack of comprehension of God's means and reasoning: the existence of moral evil, the occurrence of natural disasters, the question of the good inherent in the divine testing of human behavior and belief, the very reason for our existence. We lack so much understanding of God, and yet some would presume that the naive engineering notion of intelligent design would better reflect God's methods than does the mechanism of natural selection. One is inclined to believe that the universe continues to expand expressly to accommodate such cosmic arrogance. But again, this debate is surely irrelevant to God's compassion and purpose.

Finally, what are the personal preferences of the person-God? In chapter 1, I noted that there exists some diversity of belief about those things that are disfavored by God: perhaps certain sexual preferences, the consumption of specific foods, profit from money lending, and other activities. These might be viewed as those kinds of preferences that fall outside the rubric of Golden Rule–type morality for which we have an intuitive sense. On this matter it's my belief that God *is* the good. He is the reason for our perception of that critical asymmetry between good and evil—thus are we created in his own image. That asymmetry is written onto us in God's own hand. We may not choose to do the good all the time, or any of the time, but I believe that we know nevertheless what it is. This belief is not a dangerous invitation for a free-for-all, as some might argue, fearing it implies we each get to decide what is good. I for one feel no sense of freedom to decide. I know the distinction between my desires, my prejudices, and the good, and I'm surely not unique in this regard. So what if certain scripture could be interpreted to conflict with my notion of the good? Well, scripture is an inspiring and invaluable index to God's works, and I know that without that index I would be much the poorer, perhaps even a little lost on occasion. Yet if in rare instances there is conflict in the written word, I will opt always to trust the original. As for the probability that my *particular* view of God is the correct one? Well, it's really more a matter of faith.

I hope that in accompanying me on this unusual journey you have gained a slightly different perspective on the natures of faith and reasoned probability. If not, I would happily settle for having introduced you compellingly to the world of Bayesian inference, albeit through a somewhat unorthodox application. It's ironic to consider that the good Reverend Bayes, having been deceased for almost two and a half centuries, has by now likely had many theological uncertainties resolved for him. I wonder if the afterlife is a sort of inversion of our world when it comes to faith and probability. Perhaps there the matters to which we currently attach faith are the subject of solid, divine reason, while our terrestrial logic is viewed as being rather airy-fairy, as some consider faith here. Or perhaps this is barkingly mad speculation. For myself, I am pleased and enlightened by the numerical results of my analysis. I take intellectual comfort in the conclusion that my belief in God is not entirely faith based. Conversely, since reasoned probability takes me only partway to my belief, a role remains for faith and for the spiritual experience it creates. I'm thankful for this balance. So these are the humbly offered thoughts of a 95 percenter. But one day I think my number will be up.

Chapter Conclusions:

Faith and probability are distinctive in nature, and yet they have the propensity to influence one another.

Our model indicates that as a consequence of the disparate characters of faith and probability, faith-based beliefs can provide no legitimate basis for any form of human conflict.

My belief in God is a balance of intellectual and spiritual elements and would be greatly diminished by the absence of either.

---

appendix

The Mathematical Theologist's Spreadsheet

As I identified my evidentiary areas for Proposition G, made my assumptions, and selected my numbers, I had no doubt that you, the reader, might have taken a different path. Your assigned numbers might have been different. For that matter, you might have chosen a completely different set of evidentiary areas as your basis for determining the truth probability of Proposition G, denoted P(G). One must expect the potential for divergence of opinion in any analysis. Bayesian analysis has the advantage that it explicitly accommodates divergent opinions. I, for that matter, often wondered what might be the impact on the final value of P(G) if I had made slightly different assumptions in my analysis.

We give the name *sensitivity analysis* to the process in which we tweak the analysis assumptions to determine the effect upon the final results. For example, we might ask,

"What if the initial prior (preevidence) truth probability of Proposition G had been selected as 10 percent or 75 percent instead of 50 percent?" How would this have changed the final result of P(G) = 67%? To answer this question, we must redo the calculations starting now with each of these different prior probabilities. We would find that if the initial prior probability is chosen to be 10 percent, then the final value of P(G) changes to 18 percent; while if the initial prior probability is chosen to be 75 percent, then the final value of P(G) changes to 86 percent. Thus we can assess the sensitivity of the final result to various changes in the analysis input assumptions.

Doing repeated calculations by hand to answer these types of *what if* questions can be very tedious. A better approach is to set up a simple spreadsheet that calculates P(G) once you've specified the values of all the input numbers. In this way an input number can be changed and the spreadsheet does the work of recalculating the final result. This appendix provides instructions on how to set up just such a spreadsheet. Once you have done this, you can have hours of exhilarating enjoyment doing *what if* calculations.

These instructions are geared specifically toward the use of Microsoft Excel as your spreadsheet software. The development process for other spreadsheet software will be very similar.

1. Open a blank worksheet. Place the cursor on the column header bar and widen Column B (which will eventually contain the evidentiary area titles).

2. In Cell B2 type **Evidentiary Area**
   In Cell C2 type **D Factor**
   In Cell D2 type **P(G)**

3. In Cell B4 type **E1: The recognition of goodness**
   In Cell B5 type **E2: The existence of moral evil**
   In Cell B6 type **E3: The existence of natural evil**
   In Cell B7 type **E4: Intra-natural miracles**
   In Cell B8 type **E5: Extra-natural miracles**
   In Cell B9 type **E6: Religious experiences**
   In Cell B10 type **E7: Another area**
   In Cell B11 type **E8: Another area**

   Note that there are now two more spaces for additional evidentiary areas, denoted E7 and E8, in case you need them for your analysis.

   Figure 1 shows what the spreadsheet should look like so far.

4. Place the cursor in Cell C4, then left-click and drag the mouse down to highlight the column of cells C4 through C11. Now click on Format in the toolbar and select Cells. Select the Number tab. Highlight Number and type **1** in the Decimal places box. Click on OK. You should now be back in the worksheet.

5. Place the cursor in Cell D3, then left-click and drag the mouse down to highlight the column of cells D3

Figure 1

	A	B	C	D	E
1					
2		Evidentiary Area	D Factor	P(G)	
3					
4		E1: The recognition of goodness			
5		E2: The existence of moral evil			
6		E3: The existence of natural evil			
7		E4: Intra-natural miracles			
8		E5: Extra-natural miracles			
9		E6: Religious experiences			
10		E7: Another area			
11		E8: Another area			
12					

through D11. Now click on Format in the toolbar and select Cells. Select the Number tab. Highlight Percentage and type **0** in the Decimal places box. Click on OK. You should now be back in the worksheet.

6. In Cell D3 type **50** This should now show up as 50%. This is the initial preevidence prior truth probability of Proposition G.

7. In Cell C4 type **10**
   In Cell C5 type **0.5**
   In Cell C6 type **0.1**
   In Cell C7 type **2**
   In Cell C8 type **1**
   In Cell C9 type **2**

   The numbers in Column C are the Divine Indicator (D Factor) values for each item of evidence.

   Figure 2 shows what the spreadsheet should look like so far.

8. Now come the formulas.

   In Cell D4 type **=IF(C4=0,"",(D3*C4)/(D3*C4+1-D3))**

9. Place the cursor in Cell D4 and left-click the mouse. Now click on the Copy icon on the toolbar. Place the cursor on Cell D5, then left-click and drag the

Figure 2

	A	B	C	D	E
1					
2		Evidentiary Area	D Factor	P(G)	
3				50%	
4		E1: The recognition of goodness	10.0		
5		E2: The existence of moral evil	0.5		
6		E3: The existence of natural evil	0.1		
7		E4: Intra-natural miracles	2.0		
8		E5: Extra-natural miracles	1.0		
9		E6: Religious experiences	2.0		
10		E7: Another area			
11		E8: Another area			
12					

mouse to highlight the block of cells D5 through D11.

Now click on the Paste icon on the toolbar.

Figure 3 shows what the completed spreadsheet should look like.

And here's how to use the spreadsheet.

1. At this point, all the numbers in the spreadsheet are identical to the numbers found in chapter 7. You can change any of the input numbers, but make sure that you don't type into a cell that contains a formula. The numbers you can change are either of the following:

    a. the initial prior (preevidence) probability of Proposition G contained in Cell D3.

    b. the Divine Indicators (D Factors) in Column C for any of the evidentiary areas (see chapter 7 for an explanation of the D Factor).

2. You can add up to two more evidentiary areas by describing them in Column B and then inserting your associated D Factors in Column C. For that matter, you can overwrite the current evidentiary areas with different ones if you wish.

Figure 3

	A	B	C	D	E
1					
2		Evidentiary Area	D Factor	P(G)	
3				50%	
4		E1: The recognition of goodness	10.0	91%	
5		E2: The existence of moral evil	0.5	83%	
6		E3: The existence of natural evil	0.1	33%	
7		E4: Intra-natural miracles	2.0	50%	
8		E5: Extra-natural miracles	1.0	50%	
9		E6: Religious experiences	2.0	67%	
10		E7: Another area			
11		E8: Another area			
12					

3. Column D shows the progressing value of P(G) from row to row as the evidentiary areas are sequentially accounted for. The final value of P(G), which is the truth probability of Proposition G, is the last number in Column D, currently at 67 percent.

You are now a mathematical theologist and can do things of which Aristotle, St. Thomas, and Kant only dreamed. Please proceed responsibly.

Bibliography

Armstrong, Karen. *A History of God,* Ballantine Books (New York), 1994

The Dalai Lama. *Ethics for the New Millennium,* Riverhead Books (New York), 1999

Davies, Paul. *The Mind of God,* Touchstone (New York), 1993

Glynn, Patrick. *God: The Evidence,* Prima Publishing (Rocklin, California), 1997

Grimbol, William R., and Jeffrey R. Astrachan. *Life's Big Questions,* Alpha Books (Indianapolis), 2002

Haidt, Jonathan. "The Emotional Dog and Its Rational Tail: A Social Intuitionist Approach to Moral Judgment," *Psychological Review* 108, pp. 814–834

Hamilton, Edith. *The Greek Way*, W. W. Norton & Company (New York), 1993

James, William. *The Varieties of Religious Experience: A Study in Human Nature*, Modern Library (New York), 1994

Lindley, D. V. *Introduction to Probability and Statistics from a Bayesian Viewpoint (Part 1: Probability)*, Cambridge University Press (Cambridge, U.K.), 1965

Moyer, Doy. "A Study of Faith," *Focus*, http://www.focusmagazine.org

Muggeridge, Malcolm. *A Twentieth Century Testimony*, T. Nelson (Nashville), 1978

New Advent Catholic Encyclopedia, http://www.newadvent.org

Swinburne, Richard. *Is There a God?*, Oxford University Press (New York), 1996

Swinburne, Richard. *The Existence of God*, Oxford University Press (New York), 1991

Index

B

N

About the Author

STEPHEN D. UNWIN, Ph.D., received his doctorate in theoretical physics from the University of Manchester (U.K.) for his research in the field of quantum gravity. Dr. Unwin has written for the *New Scientist,* among other influential scientific journals. Formerly the British government's technical attaché to the U.S. Department of Energy, he is president of his own consulting firm, specializing in risk management for his various Fortune 100 clients. He lives in Columbus, Ohio.